The Shape of Knowledge

An Introduction to Paraphilosophy

The Shape of Knowledge

An Introduction to Paraphilosophy

Benjamin Davies

Winchester, UK
Washington, USA

JOHN HUNT PUBLISHING

First published by iff Books, 2023
iff Books is an imprint of John Hunt Publishing Ltd., No. 3 East Street, Alresford,
Hampshire SO24 9EE, UK
office@jhpbooks.com
www.johnhuntpublishing.com
www.iff-books.com

For distributor details and how to order please visit the 'Ordering' section on our website.

ISBN: 978 1 80341 022 7
978 1 80341 023 4 (ebook)
Library of Congress Control Number: 2022937880

A CIP catalogue record for this book is available from the British Library.

Design: Stuart Davies

UK: Printed and bound by CPI Group (UK) Ltd, Croydon, CR0 4YY
Printed in North America by CPI GPS partners

Contents

Preface **1**

Part I: On the Relation of Opposites **9**
Chapter One: The Problem with Philosophy 11
Chapter Two: Appearance & Contradiction 25
Chapter Three: The Principle of Complementarity 37
Chapter Four: The Opposites in Thinking 47
Chapter Five: Neural Bifurcation 63

Part II: On the Structure of Concepts **71**
Chapter Six: The Dialectical Matrix 73
Chapter Seven: On Knowledge 85
Chapter Eight: On Being 96
Chapter Nine: On Value 110
Chapter Ten: On Right 122
Chapter Eleven: Introduction to Syntheorology 136

Part III: On the Duality of Logic **151**
Chapter Twelve: The Laws of Thought 153
Chapter Thirteen: The Paradox of Self-Reference 163
Chapter Fourteen: The Incompleteness of Consistency 175
Chapter Fifteen: Gaps & Gluts 189
Chapter Sixteen: On Logical Consequence 205
Chapter Seventeen: Computation & Consciousness 220

Part IV: On the Shape of Knowledge **235**
Chapter Eighteen: Philosophy of Self 237
Chapter Nineteen: Sameness & Difference 257
Chapter Twenty: The Proofless Proof 277

Chapter Twenty-One: Remarks on Paraphilosophy 298
Chapter Twenty-Two: Phenomenology of Self 318
Chapter Twenty-Three: Beyond Belief 330

Notes 339

Bibliography 348

Index 361

Preface

Why This Book Exists

I never set out to become a philosopher, I don't call myself one today, and I wasn't drawn to my perspective through intensive philosophical study. Philosophy was thrust upon me, at a moment of weakness, when a crack emerged in what I thought I knew without it. This is not a typical text on philosophy, and I am very grateful that you would allow me a moment to explain why you might want to read it. I have tried my best not to overindulge in the first-person pronoun where possible in these words, but this is an introduction, so I hope you can forgive me for splurging with it here.

This book has little to say on religion, but it makes sense to start there, because I gained the first seeds of atheism at a very young age. I remember being set a task at primary school—kindergarten, that is—of drawing different kinds of animals onto a page. I remember telling an older kid that they shouldn't be drawing 'people', because people are not animals. They quickly informed me that people *are* animals, and I remember being very confused. I remember that another child in my year never attended the church services at this school, and I remember eventually realising why this was—that not everyone *believed* what we were taught was true. By the time I started secondary school—that's middle school—I had become an atheist. I did a project in my second or third year on George Carlin, and then I discovered Christopher Hitchens, and Richard Dawkins, and Stephen Fry, and I was enamoured.

I was also interested in religion in its own right. I would make long lists of the names of gods and goddesses from different pantheons. I tried correlating gods from Norse mythology with those from Greece. I even made up my own pantheon by translating words like 'fire' and 'water' into various languages. I researched the correlations between stories in the Christian bible and astronomical cycles and events; and I confronted my religious education teacher with contradictions in the life of Jesus. I was too young to acknowledge any deeper significance to the world being devoid of objective meaning. I just found it very interesting *why* people believed things we did not know as true.

This went on for some years, though I stopped spending much time thinking about religion, for atheism had just become a 'known'. I read Albert Camus' *The Myth of Sisyphus*, a book which seeks to shed light on the conflict between humankind's desire to find meaning in life, and its inability to do so. Camus concluded that our capacity to create meaning artificially, and find joy in such meaning, can compensate for its innate meaninglessness—that life might be considered meaningful if only we remain aware of the fabricated nature of the meaning we give it. This message was sufficient for me, for a while, but as I approached the end of my teens, and I struggled with the question of what fabrication of meaning would be sufficient, my perspective rapidly declined into nihilism.

I never wanted to waste time on the relative. I wanted to know what was best, and I wanted that. Insofar as there was not a 'best', I wanted nothing. I had created a problem in my mind that did not seem possible to solve—a first encounter with paradox, perhaps—and I was lost. I had always liked books, and so I naively thought that maybe I could follow in the footsteps of those atheist authors I admired. I started reading more, all while my pessimism and desperation grew larger, and I never could have predicted where this interest would lead me.

At the close of 2012, I was reading Lawrence Krauss' *A Universe from Nothing*, a book which explores how it might be possible for a meaningless universe to arise in the first place. Krauss posited that certain phenomena present in empty space could allow for the generation of our universe out of nothing, defying the cosmological conjecture—out of nothing, nothing comes. However, Krauss' idea of nothing wasn't what most people would think of when hearing the term 'nothing'; instead, it contained all of the laws of physics and the entire contents of quantum field theory. For Krauss, 'nothing' was the absence of physicality in all its forms, including those of space and time, but arguably, still, something.

Prior to this, I'd never really worried about the *before* of matter, and, rightly or wrongly, I had interpreted this as meaning that science's best guess at the origin of matter was found in something which *exists* but is not physical itself. Suddenly, flashes were going off in my mind. I started questioning everything I thought I knew, and because I had trained so long not to trust *anything* whose truth I could question, suddenly these beliefs were no good, and they collapsed. And it *all* collapsed, as though these beliefs were the scaffolding holding together my entire outlook on life. It collapsed and then there was nothing—emptiness. I knew nothing, and I knew it. I knew nothing, and I *knew* it—this thought reverberated in my mind. How can there be knowing if there is nothing?

There was nothing rational or intellectual about this feeling. I *knew,* and I could *feel* my knowing. It was an absolute, preconceptual conviction of necessity. That the emptiness is full; that nothing *means* everything; that meaning is innate; and that the absolute is a paradox that *cannot be explained,* but can nevertheless be known, because I knew it. Now, save for the flashy description I have just provided, I don't want to mystify this experience any more than it should be. I recognised that it was a psychological phenomenon that had been cultivated by my unique situation. The content of the experience is not important here, particularly for a work of philosophic intentions. The importance of the experience is the *effect* it had on me, for it is this effect that led me to write this book.

Whatever that experience was, it showed me something I did not think was possible. It showed me that, within an instant, it is possible for even the most sceptical and nihilistic of minds to *feel* absolutely certain of the importance, beauty, and perfection of existence. I report this as an account of this effect, and my life following it is evidence of its power. It was not a feeling of assurance like the one we have in response to a near incontrovertible scientific theory, nor of the certitude that 1+1=2. It was not an assurance *of* anything we can conceive, but an assurance of the existence of an assurance *greater* than anything we can conceive. It was an experience of untarnishable persuasive power, for which, I am aware, my rhetoric is not—but that is why I have written this book.

Immediately following the experience, my task was already set. There was no time to sit in amazement. There was no question that I would not spend every available moment of my life, for as long as it would take me, in finding a way to reveal this meaning. There was also no question that I would attempt to convey it *without* being able to convey it with rigour. Even if I'm ultimately wrong in all this, which is a natural and persistent doubt of the mind, nothing could stop me from risking it all, and if I should fail completely, I'll still be happy I tried.

Developing a Philosophical Science

A secondary facet of the experience that struck me was that my mind was quick to attempt to intellectualise or rationalise this 'emptiness', and as I formed thoughts *about* it, mirrored were sensations of meaning. Unable to rationally categorise the experience, I could sense successive iterations of an attempted rationalisation shuttling up towards the higher levels of my mind, and after each conceptual iteration, a sensation of meaning emerged to compensate for its incompleteness. Taken together, the sensation was that the interaction between this 'nothing' and my awareness of it, was generating an autonomously expanding duality as an attempt to relativise the paradox at the heart of experience, and that this process was the origin of my perceptions of a subjective and objective environment. This quality has been my guiding light in subsequent years, for it was a distinct quality of *complementarity*—of two interdependent constructs, who rely on each other for their own existence, and whose relation is a consequence of the actualisation of the potential.

These constructs were like two different ways of looking at the phenomenon of experience, and I fell into the habit of referring to them as 'meaning' and 'reason' respectively. By 'reason', I am referring to a discretely encoded, or mechanistic, record of experience, and something that could theoretically be written down or communicated. By 'meaning', I am referring to the continuous felt quality of experience in the immediate moment, which is never communicated, and never remembered. If reason answers the question 'How?', then meaning answers the question 'Why?'. The two viewpoints were not unfamiliar, for everything we perceive is perceived through the languages of meaning and reason; I was merely recognising them at a very basic and unrefined level. All attempts at knowledge are an attempt to acquire a reason *how* or and meaning *why*, and I felt the two as aspects of a single thing, which cannot be conceived as singular.

I realised that each view carries an aspect of the truth, that each is incomplete in isolation, and that a disproportionate adherence to either leads the individual to a biased perspective on the natural world. Meaning leads to an overly finalistic and spiritualistic view on existence, reason leads to an overly mechanistic and nihilistic view, and truth lies *in the correlation between them*. I saw my prior self to be leaning heavily towards reason, and I saw that I was ultimately no different from the theist who leans to the side of meaning. It is therefore from the view of a reconciliation between the opposites that this work begins.

My enthusiasm following this initial experience was unshakable, though I had a very long way to go to being able to express myself with clarity. I came from the side of reason, so discussing the experience on the basis of anecdote was out of the question, and I also wasn't interested in the potential psychotherapeutic effects of the recognition of meaning—I was interested in *proof*, regardless of whether proof was possible, and acknowledging that any proof would still be one side of the whole.

In 2013, I produced a booklet containing the basis of my ideas, in which I presented the thesis as a kind of pantheism, and that the problems of our time must be solved by reacknowledging the lost aspect of ourselves. Before I had even finished this text, I realised that this was not an accurate way to describe the idea, and I didn't pursue it further. I was still very young, naive, and uneducated at this time, so I needed more time to explore philosophy.

My second attempt occurred in 2016, and I knew enough of the historical discourse this time to recognise certain patterns in our theories, and the system that emerged remains unchanged in essence to this day. I released this attempt as a short film and lecture, though within a week, and after some initial positive feedback, I realised that the bold intentions of the work deserved a much better treatment than I had given it, and indeed was able to give it. I still needed more time, to learn more philosophy, so I removed the film. At the start of 2019, I would begin my final effort, researching, questioning, and expanding on everything, doing as much as I could possibly do with this mortal mind. It has been extraordinarily difficult for me, but this book is the result, released a decade following the initial event, and I'm quite content with it. I don't think I could have done much more.

The goal of the following text is to establish the foundations of a new philosophical science called 'paraphilosophy'. This is not a philosophical theory, but rather the structure of possible theories. The methodology developed to identify this structure is scientific, in the sense that it is an analysis of the ideas that philosophers have developed in our efforts to rationalise experience over the course of the history of discourse. Paraphilosophy is a consequence of the establishment of this structure as something that exists ontologically.

The text is divided into four parts. Part I discusses the general problem of opposites in philosophy and psychology; Part II develops the structure underlying our cognitions of various key concepts in philosophy; Part III extends this structure to the elementary logic underlying rational thought, and to the truth predicate itself; Part IV presents the conclusions of the inquiry in full, and particularly the self-asserting character of the structure, as well as its involvement with consciousness.

In Part I, I describe the no-progress problem of philosophy, which is that we are unable to decisively discriminate between the subjectivist and objectivist perspectives on any given philosophical problem, and that this prevents philosophy from making consensual forward movement. I then introduce the principle of complementarity, first as it is given by Niels Bohr in theoretical physics, and second as it emerges in psychology as explained by William James and Carl Jung. By using Jung's investigations into typology, I make a first presentation of the structure referred to as the 'dialectical matrix', here distinguishing four classes of intellectual temperament.

In Part II, I begin by abstracting this structure away from psychology, so that we may determine whether the same structure can model the various perspectives philosophers have developed in regard to certain basic concepts in philosophy, as well as the theories which are grounded in these perspectives. These concepts are *knowledge, being, value,* and *right,* respectively. At the end of Part II, we are able to conclude that the dialectical matrix is indeed sufficient to categorise, define, and demarcate the ideas we can come up with in regard to any of the major concepts of philosophy. I finish with an analysis of the correlations between perspectives from the same part of the structure among the four philosophic disciplines, revealing consistent themes.

In Part III, I provide a brief description of the history of logic, particularly as it relates to the foundations of mathematics. I discuss the unavoidable presence of self-reference in logic, emerging through Russell's paradox, Gödel's incompleteness theorems, and Tarski's undefinability theorem; I also discuss the requirement of non-classical logic to retain a correlation between proof and truth, as well as a transparent account of the truth predicate. I describe how the various approaches to non-classical logic can be modelled by the dialectical matrix, and end with a discussion on the relation between self-reference and consciousness, with particular focus on the work of Douglas Hofstadter.

In Part IV, I begin by re-examining a quadrant of the dialectical matrix that was previously explained to be self-contradictory, and therefore empty. At this point, we are able to determine that this perspective arises from a conflation of the subject and object in a similar manner to the set-theoretic and semantic paradoxes. Through a deeper analysis of these paradoxical concepts, we are able to identify them with self-knowledge, self-being, self-value, self-governance, and ultimately self-consciousness itself. I proceed with an examination of the dialectical matrix in light of its inclusion of this self-referential component and reveal the entire structure to be a fractal within the basic architecture of experience. The self-proving character of the possibility of self-conception, established through the act itself, and which now both includes and is included by the dialectical matrix, is the means to proving that the matrix exists as both the subject and object of thought. The remainder of the text discusses the philosophical consequences of the establishment of paraphilosophy, our new philosophical science.

I do hope that you enjoy the following text, and I hope that I have done this bold task justice.

Part I

On the Relation of Opposites

Chapter One

The Problem with Philosophy

The No-Progress Problem

To the present date, philosophy in the Western tradition has enjoyed more than 2600 years of thought and discourse. It has seen countless great minds divulge countless pages of well-considered words to its rich and devoted history. More likely than not, hundreds of millions of hours have been spent gruelling over keys and parchment alike to find those combinations of words that conform to the particular perspective of each of its great thinkers. Despite this, the pursuit of philosophical knowledge is an endeavour that has consumed more of our attention than it's paid for, for in spite of our efforts over these past three millennia, we have little real progress to show for it.

I do not mean to say that philosophic successes have not been made, nor that individuals have not made progress of their own. There have surely been plentiful occasions in which genuine understanding has shone through the darkness of mere speculative thinking. There is no doubt that knowledge has been gained; but there is also no doubt that any knowledge acquired in times gone by has ultimately failed in imparting a lasting dent in the records of history. It certainly has been insufficient to set all our philosophic squabbles to rest, for it is an undeniable fact that after these 2600 years of recorded discourse, we still tackle the same questions that bemused the ancient Greeks. It is not so much a problem of intellect as it is one of consensus, and it is the latter which shall move philosophy forward.

The individuality of philosophic opinion is a reality reflected in the attitudes of ordinary people towards the scope and hope of the discourse. For those who pass by unaffected and uninvolved, philosophy is an afterthought at best; for those who see its weakness in contrast to the strength of scientific empiricism, it is harshly criticised as hopeless or futile. Philosophy informs every choice we make, every act we sign, every word we write, but it is always a personal ordeal, for we have no greater wisdom to which we might appeal.

The academic community naturally sees things differently, for they are confronted with what appears to be progress on a regular basis. Such progress is typically a progress in theory, not a progress in absolute truth, and it is the latter we require to solidify the former into something that would resemble a science. One could walk into any university's philosophy department right now and find as much disparity of opinion as one would in its dorm rooms, and this is not to say that professors of philosophy are not especially gifted in their proclamations; academic philosophers are surely some of the brightest among us, it is rather that they have an impossible task.

There is a more systemic problem that faces philosophy: a problem that polarises and fragments our perceptions, that divides and segregates us as free-thinking individuals; a problem that renders us incapable of appreciating the positions of our fellows, and that leads us to believe wholeheartedly that our ideas outshine others. The problem devalues philosophy; it devalues the sanctity of our own opinions; it devalues the credibility of a person who would spend their time thinking about that which apparently cannot be thought about. It discredits *me*, and what this history would lead you to presume about someone like me, who is attempting to present you with ideas that you have not derived yourself.

There is not, and there has never been, a single definitive proof of any of the major epistemological, metaphysical, meta-ethical, or politico-theoretical positions, and there won't be until something is radically changed within the landscape of philosophical thinking. Even when it seems as though a theory might be destroyed, its proponents need only to sharpen their concepts, and the theory returns stronger than before. Our reasoning works well in making theories better, stronger, more consistent, but it does not work in proving these theories over others.

Despite this, it seems quite necessary that there should be proper philosophical reasons to justify certain beliefs—that slavery is wrong, that we should all have equal rights, or that reality is made of a certain kind of stuff. But our beliefs are seldom acquired from philosophy, nor are the societal changes they engender. Beliefs are caused by factors more material than reason alone, like emotion, utility, tradition, education, and societal changes occur because these factors evolve over time, and for those who are not so easily impelled to change, because of pressure from those who are. New ways of thinking are absorbed by children who grow up on the foundations of cultural innocence, wholly unaware of the quirks of their predecessors, which may be considered abhorrent in modern times. Yet, we seldom judge our ancestors; we say, 'times were different back then', as though the truth were different too.

It would be reassuring if our choices were supported by established principles, but in practice, there is always a contrary opinion, based on different ideals, and which, once we strip away our own, can be viewed as equally reasonable. Philosophy seems to express an essential paradox of the method: on one hand, there must be some theories which are basically true, but on the other, there is no reliable method for discriminating between opposing theses.

It is accordingly a crucial factor of this work to explain why philosophy behaves in this way—why the truth is so elusive, and why it's not been possible for us to put any problem to rest. These are questions that anyone who engages in philosophic contemplation is faced with, whether one acknowledges them or not. The failure of philosophy is that all too often they *are* left unacknowledged, for when a problem is unacknowledged, there's certainly no chance that it might get resolved. Perhaps we simply loathe to recognise that philosophy is stagnant because progress is impossible—that philosophical truth is simply incompatible with the human mind.

History certainly seems to lend credence to this idea, but I doubt that many philosophers would be prepared to accept it as fact. After all, what is the point of doing philosophy if it's not possible to do it well. The endeavour does seem to implicate the assumption that philosophical truth is, at the very least, structured similarly to the categories of human language, and, ideally, just as accessible as something like a mathematical formula. Nevertheless, some philosophers *have* claimed that progress is impossible due to the innate nature of truth itself, and the divergence of this nature from that of human thought.

In a 2011 paper entitled 'There Is No Progress In Philosophy', Eric Dietrich audaciously asserts that the denial that philosophy does not progress, by the majority of the philosophical community, is really a mental disorder where one refuses to accept an obvious reality—a conclusion I should clarify I don't share.[1] Nevertheless, Dietrich does give a good account of the no-progress problem in philosophy, and of particular relevance to the present work is the perspective of Thomas Nagel, as one of two explanations Dietrich gives as to why philosophy is stricken with such a problem.

For Nagel, progress in respect to certain philosophical problems is unattainable because there are invariably two coherent yet contradictory perspectives through which one can approach that problem: one, from a principle-based, first-person, subjective standpoint; and the other from a fact-based, third-person, objective one. Nagel expounds that the bifurcation of perspective makes insolvability an inherent feature of the questions themselves, so that we could not, no matter how much intelligence we employed, find for them a single satisfactory answer.[2]

The duality between subject and object-based viewpoints is a universal feature of the questions we call philosophic. The issue that philosophers face is not merely that two such perspectives exist, but that the conclusions of each stand in direct opposition to the other. When we think about the world in respect to the subjective, for example, we acknowledge that our only access to reality is dependent on our own experience of it. From the perspective of the subject, we are intimately connected to reality and play a role in its perpetuation. In this sense, reality appears as something we are *doing*—that is dependent on consciousness.

On the other hand, when we think about the world in respect of the object alone, reality appears to be unfolding purely in accordance with natural laws that are entirely disconnected from, and unconcerned with, anything that might be conscious of them. In this sense, reality is something that is happening *to* us, and our experience is merely an evolutionary mechanism by which we become able of perceiving and interacting with it. We are resigned to this endless war between contradictory viewpoints, and any effort to resolve this conflict demands an assumption that the conflict is indeed resolvable.

The present work begins from this assumption but operates with a very different view on the nature of no-progress in philosophy. On this view, our efforts to discriminate between opposing perspectives presupposes certain ideas concerning the nature of truth, and that these assumptions prevent us from undertaking philosophy with an open and unbiased mind. To boot, our assumptions regarding the nature of truth force us into pursuing a specific approach to philosophy, and this approach cannot arrive at solutions no matter how hard it is tried. On this view, *we* have been dictating to philosophy what kind of truths we would be willing to accept, devising theories that fit these specifications. Consequently, philosophical theories concern only the mental states of their devisors and have no claim as absolute descriptions of reality. On this view, we have a limited conception of truth because there is an incompatibility between our ability to *intuit* the necessity of truth, and our ability to recall and define what that truth actually is.

The Assumption of Monoletheism

The basic assumption surrounding the nature of truth is that the state of affairs relating to a given philosophical problem corresponds to some consistent set of ideas or some particular theory. It is then reasonable that the reality existing beyond our dualistic and fragmented perceptions conforms to the same rules of inference that allow a subject to become aware of an object within it. The basic assumption is so prevalent throughout the history of discourse that we do not recognise it by a particular name, or as a theory. Naturally, any belief which is accepted as true by default does not need a name, for we envisage no philosophy existing without it. The name I have used in private over these past years is *monoletheism,* from the Greek *mono,* meaning 'one', and *aletheia,* meaning 'truth'—'one truth'. This term is quite natural, and I am likely not the first to use it.

Monoletheism is a view on the nature of truth in relation to philosophy. It states that mutually exclusive philosophical concepts, and theories, cannot be true simultaneously, and has been an unquestioned and unacknowledged supposition of nearly all of Western philosophy since the time of its inception. Its use has gone unnoticed because we have also assumed that its denial would reduce philosophy, either to a complete refusal of truth in the form of extreme scepticism, or to trivialism, by exploding all propositions to be regarded as true. It is easy to see why monoletheism has consumed philosophy so omnipotently, and it would certainly require a pretty good reason as to why it should now be abandoned.

Of course, monoletheism is not merely an intellectual and methodological precept, but a biological and physical one too. Our environment is presented to us monoletheically; our brains process information monoletheically; and our ability to communicate, to discriminate, and to act with intention is dependent on the veracity of some things being right and their opposites being wrong. We have monoletheism programmed into us, and for purely practical, evolutionary reasons. Natural selection cares not for our access to the transcendental, and there is no guarantee that the same traits which produce thoughts for survival might also produce thoughts for knowledge. We see in a way that makes sense to us because *we must* see in a way that makes sense to us; evolution magnifies that which makes a difference *physically* and leaves out that which does not. We wouldn't get very far, for example, in trying to play a video game by looking directly at the code, and if our reason is shaped by the same forces as our environment, it could in fact be a detriment for us to see things as they actually are.

Human beings have proven themselves decidedly capable of gaining knowledge of the phenomenal, which we can be confident is conditioned by factors familiar to the intellect. But quantum theory already describes a world that looks quite unlike the world of experience, and so monoletheism is not merely a belief in a lack of contradictions, but a belief that our capacity for empiricism, linguistics, and analysis in science is not at odds with the structure of the real.

To employ monoletheism in philosophy is to assume that there is one true logic, and that this logic is shared between both science and philosophy. The predominance of monoletheism is not purely a consequence of the monoletheic nature of the object, but also of the historical marriage between science and philosophy, and the lack of a clear ontological distinction between the scope of each. This conflation is not extended to subjects like psychology, where we acknowledge the role of subjectivity as essential, and where we do not demand for every psychological fact to apply to every psychological subject. It is therefore important to understand how monoletheism has been embedded into philosophy from its start, and how it has not been refigured following its divorce with natural science.

Natural Philosophy

When Western philosophy was birthed around 600 years BCE in ancient Greece, its earliest practitioners were referred to as the *physikoi*, meaning 'physicists'. They were called this because they were among the first in the Western tradition to have rejected the prevailing reliance on mythology and theology when attempting to understand natural phenomena. They were *physicalists*, and they were developing the earliest form of what we now know as physical science.

The physikoi did not start from a position of scepticism about the phenomenal world, for they sought explanations based on natural elements, and they were not primarily concerned with the nature of knowledge for they sought a departure from supernaturalism and superstition. The physikoi were naturalists, and so adopted the assumption that reality is objective and consistent, that we have an ability to perceive reality accurately, and that rational explanations for elements of the world do indeed exist.

It was not until 200 years ago that philosophy and physical science truly became disentangled. As the considerations of philosophy grew larger, *natural* philosophy became one of its branches, and only when specialisation became more commonplace did 'natural philosophy' simply become 'physics'. The development of both Western philosophy and physical science have been fused together as a single inquiry for the great part of their development, following in the tradition of the ancient physikoi, the first scientists, but also the first Greek philosophers.

Different approaches to the study of knowledge were distinguished a short time later in the works of Plato and Aristotle. Aristotle reemphasised the importance of sense perception for philosophical thought, and he also developed the first formal system of logic. He is also the specific individual to whom we can trace back the origins of monoletheism, and the traditional source for the law of non-contradiction. This law was an important element of language long before Aristotle, but he was the first to formally cement its importance for the logic of philosophy.

Aristotle's law of non-contradiction asserts that contradictory propositions cannot be true simultaneously, that nothing can exist in opposing states, and that no person can believe a thing to both be and not be at the same time. These principles form the basis of what I refer to as monoletheism, and they are implications of a basic statement concerning the nature of truth. That is, contradictory claims are mutually exclusive, and every decidable proposition has exactly one of two truth values, those being 'true' and 'false'.

Aristotle's development of logic was a great step forward for both rationalistic and empirical inquiries into nature; indeed, the law of non-contradiction is essential for maintaining the consistency of philosophical positions. The point I would like to make here is not that there is anything wrong with the principle, but that it was implemented in philosophy with one eye on sense perception and observation. Aristotle was a philosopher, but much of his surviving writings concerns subjects like biology, astronomy, geology, and physics, and there was no notion that physical science and philosophy are different subjects potentially operating in accordance with different logical rules.

If Western philosophy had risen in a different time and a different place, it may never have shared its evolution with physical science. Indeed, the ancient and venerable philosophies delivered from the Eastern tradition, such as those of Laozi or Nāgārjuna, did not develop alongside the rigorous empiricism of physical science, and perhaps this is evidenced by their divergence from Aristotle's principle.

There *was* one Presocratic who also contradicted Aristotle's logic: the mysterious autodidact Heraclitus. Heraclitus was called 'The Obscure' by his contemporaries on account of the paradoxical nature of his claims, and he is often cited as someone who denied the law of non-contradiction. Heraclitus is famous for the aphorism, "Into the same rivers we step and do not step, we are and are not,"[3] and it was partly in response to the contradictions imposed by Heraclitus that Aristotle dubbed the law of non-contradiction as "the most certain of all principles".[4]

For Heraclitus, the transition between complementary properties gives rise to our apparently stable reality, and the fundamental nature of this reality is itself that of a war between opposites—between harmony and strife, stability and flux. "The only constant in life is change," is another aphorism attributed to Heraclitus, and points towards an underlying unity between the idea of a thing becoming different and the idea of it staying the same. The only absolute truth is that there is no absolute truth, and so Heraclitus would have us strive for a dynamic understanding of nature, in accordance with the *logos* by which it unfolds, rather than for a static system of justified beliefs. Heraclitus' perspective was unique among the Greeks and following him seldom have thinkers taken up where he left off. It was Aristotle's principle that would guide philosophy throughout the centuries, as the law of non-contradiction has reigned supreme ever since.

The Reign of the Object

Today, it is quite clear that physical science and philosophy are vastly difference subjects. For one, physical science moves forwards linearly, while philosophy only swells. Both disciplines develop theories, but the former has access to the domain of its theories, and so, through collaboration with the empiric, scientific theories are constrained. Over time, we converge upon a single line of thought, and, in this way, theoretical science is guided by fact, just as empirical science is guided by theory, and it does not alter our results to consider the subjective processes involved in developing either.

If philosophy is looked down upon at all, it is through its comparison with science; for we see that science works, and we see that philosophy does not. But work for what? Perhaps philosophy *is* working, and it's the comparison with science that is broken. If we should retain the same assumptions for both disciplines, it is natural to conclude that philosophy is futile, that we should accept what science can teach us and not strive for anything more, and that knowledge can only ever be of the phenomenal variety, and not the transcendental.

Acknowledgement of science's power to create knowledge, and philosophy's powerlessness in doing so, is engrained into the history of our discourse. It has not been a smooth path, but on the whole, we can see that philosophers are increasingly impelled to utilise the rigour and logic that leads empirical science into new knowledge, so that we may maintain some authority of the craft. The continental tradition is a slight kink in this path, but with the advent of analytic philosophy in the 20th century, the formalism and reductionism that facilitates science has dominated philosophy as well. Science has led the charge in organising the chaos of experience into fragments of information that serve great material purpose and function. If philosophy has wanted to retain its credibility, and not fall back towards the inane and incoherent, it has had to observe, within its courses and in its journals, the rigorous methodology that was finding such great success in science.

Analytic philosophy is an approach to gaining knowledge, and any approach to gaining knowledge with hopes of completion should be structurally unconditioned by the kind of knowledge it could possibly gain. Otherwise, we should be studying of the colour of objects while wearing coloured glasses, and our methods should influence our results. There is a particular perspective most compatible with analytic philosophy, and much like physical science, it is a perspective that is blind to the whole. It is a perspective that relies on the reality of the object, and like physical science, it must reduce the subject in order to work.

In the 2020 PhilPapers Survey it was found that 78% of academic philosophers were confident that the objects of experience exist entirely independently of our perception and awareness of them, and 71% that the reality described by science, including those elements we don't perceive, is objective and not symbolic.[5] The same survey found 87% of philosophers to believe we have philosophical knowledge already, and 71% that a violation of monoletheism is impossible in principle. If there were not strong theoretical opposition to these ideas within the discourse, these figures might easily be mistaken for consensus.

Our experience appears to flourish out of the object; we can *see* the objective world, and our living depends on it. Our dependence on objective factors for survival, over the course of evolution, has surely sharpened the mechanism by which we represent our world, and purely to make it clearer, crisper, more solid and more real, so that we might interact with what really exists more reliably. What really exists for our world *could* be a direct concrete replica of experience, but it could just as well be numbers in a computer program, and sense perception might never know the difference.

We are biologically conditioned to the object, and the same facilities by which we perceive it permit our effectiveness at science. If the most extreme objectivist theories in philosophy are true, then science shall be all we need, and philosophy may dissolve into history. Under the strongest objectivist perspective, everything adheres to mechanistic principles and is reducible into its parts. There would be no free will, moral concepts would exist purely for their utility, and we would no longer need to solve the mystery of consciousness, for consciousness, as such, would be merely an effect of the brain.

Yet, in spite of our devotion and specialisation towards the outside world, there will always be those who remain receptive to the internal representations present within the psyche, and to these individuals the inside world is as palpable and real as the world they see outside of them. The realm of the subject is going nowhere from philosophy, and no amount of empirical knowledge will prevent us from feeling intimately that there is significance to our experience—that there is truth to be found where science has no reach. The major limitation of objectivist theories is that their attractiveness is purely subjective. It is the inescapability of subjectivity that prevents an objectivist victory, and it is the power of the subject in opposition to the object that prevents progress from our discourse. Our *only* access to the world is via the mind, and so our oppositions imply each other just as objects imply a subject.

Why consciousness would evolve without any function is a question that remains to be answered, and if it does have a function then the problem is only made harder. Unless we understand our ability to learn what the world is like, there is no sense in calling it objective or otherwise. We must extrapolate *from* our ability to objectify our ability to know, and any limitation of this ability is a limitation of the strength of our arguments. We can describe the past by observing the present, but inasmuch as it concerns our immediate experience, the primary existent is our ability to perceive, with the objects of perception coming after it.

In the following pages, I hope to illustrate that the no-progress problem of philosophy implies and is implied by the tenets of monoletheic thinking, and that once we replace this paradigm with one that works, the progress we have made in our theories will become genuine progress towards truth. In order to achieve this goal, we will be required to deconstruct what it means to ask a philosophical question; we will need to reinterpret our own capacity for knowledge; we will need to transform the basic logic that guides our philosophical considerations. In short, we will need to adopt a new kind of common sense; but this *is* all we'll need, and this new common sense is waiting anxiously to be found.

Chapter Two

Appearance & Contradiction

Naive Realism

It is quite natural that there should be some kind of structure to the ideas we come up with in philosophy, because there is 'some kind of structure' to our experience as a whole, and our ideas about experience are bound to reflect this structure. The duality between subjective intellectualisation and objective representation is the most immediate structure of our experience, and it is this same duality that provides the foundation for the general division of perspectives found in our discourse. Despite this, the presence of duality within our perception has generally not been taken as grounds to conclude that what we perceive comprises a duality also.

True metaphysical dualism, of the sort espoused by René Descartes, which asserts the existence of both mental and physical substance, is victim to a difficult problem. If the mind and body are made of distinct substances, then the activities of those substances are independent. There may be a change in physical properties without a corresponding change in the mind, and possibly vice versa. The question arises: how do mental states impart change, or influence in any way, physical states such as those in brains? If they do not, why do mental states exist at all, and how did they emerge in the world via evolution? As Karl Popper demarcated for the physicalist ontology, "Physical processes can be explained and understood, and must be explained and understood, entirely in terms of physical theories."[6] The physical world is a closed system, and this leaves little room for mental causation, at least if those mental states are not simply reducible to, and supervene upon, physical states in the brain.

This is but one of several reasons why the notion of an ideal substance, existing apart from and in addition to the physical, is not particularly popular in modern times. Nevertheless, the fact that we talk about things called 'mental states' means there is something it is like to have them. If these states are then asserted to be essentially unreal, then *there* is a very obvious way in which our perception is not mirrored by reality: we perceive mind and matter as fundamentally different but deny that they actually are. The difficulty this divergence creates for philosophy cannot be understated, for if our perceptions are not reflections of reality, how can we ever know what is true *beyond* perception? Moreover, how can we trust the objectivity of *any* experience?

If we are to conclude that the appearance of mentality is an illusion, and that our experience of matter is at root material itself, then it follows that appearances are deceiving. However, if we assert that our access to the physical world, which is to say, mental states, are illusions, then how can we be sure that the appearance of a physical world is not illusory too? We operate under the presumption that there *is* a world out there, and insofar as we operate successfully it seems that this presumption is true. If it *is* true, then we know it only because our mental states reflect it, and the matter of *how* mental states reflect an outside world is very much the problem needing to be solved.

The quite natural idea that the mind is a mirror of the external world, and that what we see and sense is, quite simply, what really exists outside of us, is a favourable view because it means that we can gain reliable scientific knowledge by simply observing what the world is like. It has been several hundreds of years since this was the received view in the philosophies of science and perception, however, and not just because there are properties required to express scientific theories which are not directly perceived. It is also because there are properties of our perception that are *not* required to describe the objective world, and this implies an explanatory gap between the realm of science and the realm of external perception.

Naive realism, also referred to as common-sense realism or direct realism, was famously attacked by the Scottish philosopher David Hume in the mid-1700s. Hume was a naturalist, and an empiricist to boot, but his scepticism towards the reliability of sense perception led him to the claim that empiricism can never justify the belief that our reason actually mirrors what the world is like. That is to say, concepts which are necessary to our understanding of natural phenomena, such as the idea that any one event is caused by another which proceeds it, do not mirror real features of the world.[7]

Sense perception, Hume argued, shows us what has happened, but not what will happen, and though we may predict the future, we cannot know it until it's here. The idea that all events have causes is therefore subjective, not a part of the natural world, and any employment of causal inferences is not justified by empirical science. We cannot gain universal knowledge by observing probabilities, and any extrapolation from specific events to general laws cannot be rational. We cannot conclude, for instance, that all swans are white just because we have only ever seen white swans, and we cannot assume that the laws by which nature appears to be unfolding will be persistent into our future, just because they have always held in our past.

Hume's argument reiterated a perspective that became popular in the fourth century BCE, originating with a Greek sceptic called Pyrrho of Elis. The Pyrrhonists saw that all conclusions reached via inductive reasoning, which is to say, that are based on evidence, are victim to circularity since the only way that we could justify the validity of inductive reasoning would be by using induction itself.[8] Hume utilised this line of thinking to argue that the reliability of science is necessarily unscientific, or rather, that science cannot be justified via science.[9]

The distinction between scientific and perceptual reality can be understood in terms of what the English philosopher John Locke referred to as primary and secondary properties. The former, such as shape and motion, are said to exist within the object, while the latter, such as colour or smell, exist only within the mind. We could define naive realism as the belief that qualities like colour and smell are just as objective as physical properties like shape and motion. However, Hume saw the distinction between primary and secondary properties to be too vague, for the former are ultimately extrapolated from the latter, and even if the distinction is valid, it would be impossible to determine which properties are innate in the object, and which are products of the biological mechanism.[10]

By the 18th century, it was widely acknowledged that our perception is not a direct reflection of reality but a representation of the same. For Hume, this was an important recognition for it led unavoidably into scepticism towards the possibility of gaining reliable knowledge of the external world. This is, again, because it would be impossible for us to know whether the results of our observations are fundamental properties of the object, or merely representations created by the mind. This was a reluctant conclusion that, as an empiricist, Hume sincerely wanted to resolve.

Transcendentalism

Ultimately, Hume did not succeed in resolving the problem of realism, and the task was left to the German philosopher Immanuel Kant who followed him. Kant sought to save science from scepticism by reasoning that the fact our minds do not mirror the real world does not make knowledge unobtainable, because the *real* world, and the world of appearances within it, are two entirely different things. It is the world of representation—

the sensible world—which is the true object of science, not the world as it is beyond experience. In other words, that which we call nature is precisely how reality appears in experience, while the experience-transcendent reality is *never* the object of empirical inquiry. Science is not affected by the inaccessibility of the real world, because that's not what science ever sought to describe in the first place.

It's easy to see how something like colour is a property of subjectivity, but Kant was suggesting that *all* perceived properties are subjective properties also, and that the entire domain of science is conditioned by subjectivity. Kant proclaimed that our knowledge conforms not to what the world is actually like, but to what *we* are like, such that the way that reality appears to us conforms to the structure of our own ability to perceive it. He posited that features we naively consider to be essential to the nature of reality, such as those of space, time, and causality, are applied by the act of experiencing a transcendent reality, which is itself timeless and nonspatial. In other words, space and time are properties of *us*, not properties of the perception-transcendent, or *noumenal*, world.

Kant believed that his work expressed a Copernican revolution in philosophy, for just as Copernicus had demonstrated that it is the earth that revolves around the sun, so Kant proclaimed it is the objects of sense perception that conform "to the constitution of our faculty of intuition".[11] He saw it to be our shared psychological structure which creates conformity in our experience, and that this conformity is the basis of empirical science. It is in this way that Kant saved science from Hume's scepticism, because he enabled us to directly reconnect our perceptions of phenomena with the objective world, making any empirical inquiry a reliable account of nature.

While Hume had argued that our subjectivity was so tangled up with nature that we flounder to gain an objective grasp of it, Kant argued that we *can* develop a form of knowledge which works just as if we were not tangled up with it, given that this entanglement is precisely what we are investigating. In order to save science, therefore, according to this idea, we are required to accept that nature is adapted from some more fundamental reality, and that the properties we derive of the world express relationships between this reality and our own subjectivity.

However, though we have intuition about that which we have contributed towards the appearance of the world, we cannot extend our reason towards the transcendental. If we do attempt to reason beyond our perception, eventually we find ourselves to be confronted with contradictions. These contradictions arise from our ability to look at the world in two different ways, the affirmation and rejection of each being equally reasonable. For example, in accordance with the laws of nature, every event must have a cause, but if every event has a cause, then there can be no first cause, and if there is no first cause, then there is no cause by which events are ultimately determined. As a second example, everything must be composed of basic parts, but there are no basic parts to be found. Kant saw each of these contradictions to be a "natural and inevitable illusion" derived from an improper use of concepts, and evidence of the division between the world of phenomena and the world-in-itself.[12]

Kant's supposed Copernican revolution was incredibly influential to those who followed him, and his work set the stage for much of the following century of discourse. However, some of Kant's direct successors, though they accepted his assertion of the inevitability of contradiction, disagreed with his conclusion of a transcendental idealism. For Kant, the conclusion was necessary, for without the transcendental, we would have a representation but nothing to represent. It is that Kant could not fathom the contradictions to reflect anything actual in the world that reason must fail when extending beyond the field of perception.

The Dialectic

Following Kant, another German philosopher named Georg Hegel favoured a more radical solution, for Hegel took the contradictions inherent in human reason not as cause to doubt it, but as evidence that contradiction is essential to the nature of the object. He saw that "every actual thing involves a coexistence of opposed elements",[13] and so we *can* gain knowledge of the unperceived because the same principle that makes *reason* dialectical informs the development and expression of the world itself.

For Kant, the dialectical, contradiction producing, nature of reason is subordinate and inferior to the understanding, which operates in accordance with monoletheism. For Hegel, the dialectical nature of reason is the fuel of understanding, which assimilates the contradictions into more general concepts. "What is rational is actual; and what is actual is rational."[14] The dialectical nature of reason is the very essence of reality, and the common element between the mind and the world, which is required for our capacity for knowledge.

The term 'dialectic' literally refers to a conversation between parties in which each side holds an opposing opinion. It was first introduced to the Western canon in Plato's Socratic dialogues, where it was used as a method of dismantling a certain belief or theory in showing it led to an impossible conclusion. Classical dialectic was a predominantly negative process that sought to collapse an idea from the inside, often without offering another alternative. In Plato's dialogues, this is achieved via a series of increasingly challenging questions offered by Socrates to his interlocutors, guiding them to the ultimate conclusions of their arguments, which was always that of contradiction.

Hegel's method differs from the classical approach as he saw that dialectic could be used for more constructive purposes. For Hegel, opposing philosophical concepts, such as those of 'subject' and 'object', are implicitly contained within each other and give way to each other when pushed to extremes. Accordingly, we cannot call an idea or belief false insofar as it is contradictory; we must rather seek an understanding that encompasses both the belief *and* its contradiction. Whereas Kant's monoletheism saw antinomy to lead unavoidably to the impossibility of knowledge, Hegel, insofar as contradiction is present in the concepts themselves, saw that dialectic could be used to generate new and more complete ideas.

The way that we achieve such forward movement is via *sublation,* which is actually a distinctly negative process. It is through the self-reflective or 'absolute' negativity of the act of negation that sublation is able to produce positivity. That is, the negation of an idea is simply that which the idea is not—it is negative—but the negation of what an idea is not is *non*-negative. This idea of using double negatives can indirectly define and assert what a concept is positively, in terms of what it is not. As such, the sublated concept *is* not what it is not.

By continuously sublating or assimilating the negation of a concept into the negation of its negation, Hegel believed that he could show such concepts to contain their opposites within them, and consequently that they are innately contradictory. Each negation is negated, and the result is negated again. The process ensures that nothing is ever lost within the ongoing sublation, and the resultant structure accumulates positivity and being. Hegel viewed the entire history of thought *and* society to be based in dialectic; contradictions arise on one level and are overcome at the next, the struggle between opposites being the major driving force for positive change. In Kantian terminology, each stage can be thought of as a synthesis between a thesis and its antithesis.

The process of sublation ensures that the entire dialectical structure is filled with negation, and so the contradictions of reason are not resolved by ascribing them to separate perspectives. For Hegel, contradiction does not simply exist *between* opposing concepts, but is a part of the concepts themselves. Thus, even when we sublate concepts, and reach a higher perspective, the resultant notion will always reveal itself to have its own contradiction, implicit within it, which must be sublated via further negation. For instance, free will and determinism are not simply to be regarded as equally true but in different respects; it is rather that each concept *becomes* the other and must be grasped with its other within an infinite spiral of dialectical activity. Every cognition we produce of the identity of this spiral will be replaced by higher forms throughout the passage of time, as it homes in on the Absolute concept.

Contradiction is the fuel of Hegel's logic, and that which allows him to resist Kant's distinction between phenomena and noumena, replacing it with an evolving entanglement in which conflicts come together. Hegel's philosophy is entirely juxtaposed to the modern analytic trend of fragmentation and reductionism, and also to the tradition leading back to Aristotle that preceded him. Instead, Hegel saw all things to be parts of a greater whole, and that an understanding of this whole can only be reached via the integration, not the separation, of its parts.

Hegel's philosophy is a pillar that has stood alone in its confrontation with monoletheism, though it is not one without controversy. For one, since all concepts are built of the successive negations of previous cognitions, there is no obvious reason why the dialectical process should not go on forever, never arriving at some concrete knowledge. Hegel was quite sure, however, that the dialectic would end in some *absolute* perspective, encompassing everything beneath it and only describable with contradictory language. Perhaps it is the case that at some point further negations of our cognitions will be identical to each other, revealing the unity between an absolute end and an eternal process, and validating Hegel's predictions.

The major complaint to be made of Hegel by analytic philosophers of course is his disregard of classical logic, and his deconstruction of the consistency of philosophy. In the *Science of Logic*, Hegel denies Aristotle's law of non-contradiction when he says, "The something which ought to have been either +A or –A is here attached to the +A as well as the –A," and, "All things are in themselves contradictory."[15] It is no surprise that Hegel's violation of this law of thought is seen by many to discredit his system to incoherence, even if this judgement is ultimately undeserving. Classical logic had stood as the incorruptible and inextirpable edifice of rigour and reason for over two thousand years, and to challenge this custom was always going to be met with violent opposition. Bertrand Russell, for instance, in his *History of Western Philosophy*, concluded his section on Hegel by stating that Hegel's work "illustrates an important truth, namely, that the worse your logic, the more interesting the consequences to which it gives rise".[16]

There is more to say on Russell's relation to Hegel, for it was partly in response to the obscurity of Hegelianism, which, as Russell saw it, had begun to pollute the minds of British thinking, that he and G.E. Moore set out to develop a new brand of philosophy grounded firmly in the reductive clarity of logical and linguistic analysis. By the turn of the 20th century, analytic philosophy was already emerging as the leading philosophic tradition, and the brief influence of continental philosophy, which began with Kant's response to Hume's scepticism, was swiftly pushed to the side; scientism would reign supreme again.

The Unity of Opposites

One thing Hegel was unclear on was that contradiction is not the only way that we can characterise opposition, for there is another kind that is present in a wide variety of qualities and concepts. For example, 'pushing' and 'pulling' can be contradictory when the resultant forces are in opposing directions, but they can be complementary when they are in the same. We might say that Hegel's view was to take 'pushing' and 'pulling' in two directions to result in a kind of sideways motion, but there is another way for opposing forces to produce positive change when those forces are working to *assist* each other, rather than to hinder. Moreover, there are opposites that ascribe the same object with two different perspectives—something might be helpful in one sense while harmful in another—and these perspectives need not contradict each other.

Many of the oppositions that Hegel enumerates are not contradictory in a logical sense but are rather dichotomies that have opposing meanings. Dichotomies and negations are not necessarily the same, yet Hegel paints them with the same dialectical brush. This again reveals the underlying current of Hegel's thought as that of a conflict between opposites which must be overcome. As Slavoj Žižek puts it, Hegel's method was to "demonstrate how every phenomenon, everything that happens, fails in its own way, implies a crack, antagonism, imbalance in its very heart".[17]

In a complementary opposition, two concepts are distinct but reciprocal. There is no need for sublation, assimilation, or reconciliation because the concepts are not conflicting; they are sufficient in their natural form and need not be transformed into more general cognitions. While the traditional, non-Hegelian understanding of contradiction leads to a dismissal of one or both sides, complementarity leads to their simultaneous acceptance. There is therefore a clear dialectic at play between these two approaches to opposition, for whereas contradiction leads naturally into strife and disharmony, expressed by the notion of a war between opposites, complementarity cultivates coexistence and harmony between allies.

There is no clearer concrete manifestation of the relation between opposites than that which is revealed in society, where the distinction between competition and cooperation reflects the distinction between contradiction and complementarity respectively. Contradiction-based thinking, which is to say, the kind that discriminates on the basis of difference, is the root of many problems we face in modern social life. Distinguishing between opposites is a harbinger of fear, for if you are fundamentally different from something else, then there is always the possibility that the other will encroach upon your own power and freedom. On the other hand, contradiction-based thinking has significant advantages, and is a major producer of progress. It is not simply fear that drives people to compete, but also the opportunity for growth.

For two things to be antagonistic, there must also be a sense in which they are similar, lest we have no cause to compare them at all. Hegel's dialectic omits any agreement between opposites and therefore finds no value in them *as opposites*. Whereas contradiction-based dialectic seeks to resolve conflict by destroying both perspectives and creating a compromising third, complementarity-based *dialogic* seeks to find a way for opposing perspectives to coexist together. A science of the unity of opposites must recognise both of these approaches as forming precisely the kind of opposition it seeks to unify, whereas Hegel only seems to care about one. To realise the identity in difference, we must acknowledge the difference in identity.

Chapter Three

The Principle of Complementarity

On the Nature of Light

The philosophy of Hegel and other early 19th century idealists like Fichte and Schelling, and of course also Kant before them, emerged as the philosophic appendage of a much wider response to the mechanism and reductionism betrayed by Enlightenment period rationalism. German idealism would outlast the greater movement of Romanticism, but it started to decline at the turn of the 20th century with the inauguration of analytic philosophy and the restored dominion of scientific positivism. Around the same time as Hegel was conceiving of his contradiction-based approach to the problem of opposites, however, a new tradition in physics was being born that would go on not only to reveal the limitations of Newtonian mechanics but would also eventuate in the rediscovery of the problem of opposites, only this time from the perspective of complementarity.

The 'duality paradox' would take full form in physics around the same time that Hegelianism met its final demise in Britain, and in looking back now it seems that the dialectical baton was passed from philosophy over to science, championed by a Danish physicist by the name of Niels Bohr. Here, the debate would concern two opposing conceptions of the nature of matter, a dialectic Hegel alluded to when he said, "It is said that matter must be clearly either continuous or divisible into points, but in reality it has both of these qualities."[18]

Since the 1600s, physicists, or natural philosophers, had been debating as to what is the best way to provide a description of light. Many took observed phenomena such as diffraction and refraction to mean that light behaved as a kind of wave. In 1637, René Descartes presented one of the first arguments in favour of the wave-image of light, and in 1690, Christiaan Huygens formulated its first detailed theory, where it was suggested that light was a disturbance in an underlying physical medium called the 'luminiferous ether'. That is, a wave is not an independent entity, but rather a fluctuation in some other substance. A wave in the ocean is a disturbance of water, sound is a disturbance of air, and so it was taken for granted that for light to be a wave there must be something grounding it which is doing the waving.

Not everyone agreed with this view, however, and some natural philosophers of the time favoured a particulate interpretation of the nature of light. Chief among these was a young Isaac Newton, who in the 1660s discovered that white light was composed of several different colours; he developed the hypothesis that these different colours were due to the slightly different masses of light-particles, such that heavier particles would travel slower through a transparent medium, explaining the differences in their refractive indices. Despite the apparent rationality behind the wave-view of light, it was ultimately Newton's particle-view that emerged victorious in the 18th century, and Newton's reputation may well have played a role in this.

The story continues around 100 years later, when an Englishman named Thomas Young devised an experiment that would demonstrate light to exhibit distinctly wave-like behaviours. By directing a ray of sunlight through two pinholes, creating two beams of light, Young discovered that the beams interfered with each other in a manner that is characteristic of waves. If the light is shone through these slits and onto a screen behind it, the interference pattern is visible as a series of bright and dark regions. Young rightly took this to be caused by the constructive and deconstructive interference of the two waves: where two crests meet, a bright area results, and where two troughs meet, a dark one. This effect is known to occur with interfering waves, and with Young's findings the particle-view had been thrown into serious doubt. Young was also able to correct Newton's theory of optical colour with an explanation of the varying wavelengths constituting the light. It was thus the wave-view that would be at the forefront of physics as it entered the 19th century.

Young's findings were eventually given a solid theoretical foundation in 1865, when James Clerk Maxwell developed a series of equations that would unify two separate phenomena—electricity and magnetism.[19] Maxwell's resultant formulation of electromagnetism described light as a propagating transverse wave within underlying electric and magnetic fields. The final nail in the coffin, for the particle-view, was sunk in 1888, when Heinrich Hertz successfully produced the first radio waves. These waves were found to travel at the same speed as visible light, and undergo similar processes of reflection, refraction, and diffraction. In combination with Maxwell's theory of electromagnetism, we could now accurately describe light as an electromagnetic spectrum containing many different wavelengths. Only a fragment of this spectrum is visible to the human eye, but there is otherwise no difference between the light emitted from a bulb, the light emitted from a radio transmitter, or light of any other wavelength.

With Maxwell's theoretical and Hertz's experimental evidence in favour of the wave-theory of light, by the end of the 19th century the case had largely been closed; it was indisputably concluded that light was indeed a wave. However, while it seemed that the story of light, and indeed the entire field of classical mechanics, was beginning to reach its end, there were still some problems that remained unsolved. As it turned out, to solve these problems would be to unlock a new era in the history of physics.

One of these problems had to do with something called 'blackbody radiation', which, to give a simple description, refers to the thermal energy emitted from a heated object. The issue was that Maxwell's theory of electromagnetism seemed to be inaccurate, via the Rayleigh-Jeans law, in calculating the radiation that would be emitted from an object at high frequencies. In fact, the law predicted an intensity that approached infinity as the wavelength grew smaller—a result that was not only inconsistent with experimental findings but was also beyond the scope of physical possibility.

In the year 1900, Max Planck was able to show that if the radiant energy emitted from a heated object was discontinuous, then a value could be calculated that matched empirical findings. Then, in 1905, a 26-year-old Albert Einstein provided Planck's calculation with a physical description of light as composed of discrete packets called 'photons'. Einstein's depiction of light as a beam of particles is expressed in his solution to the photoelectric effect—another phenomenon that conflicted with the wave-theory of light.

The photoelectric effect occurs when light is shone, for example, on a metal surface, resulting in the ejection of electrons from that surface. Electromagnetism predicts that the kinetic energy of particles emitted should be proportional to the intensity of light shone upon them, and that the quantity of particles emitted should be inversely proportional to the wavelength, but this was not the case. Instead, precisely the opposite situation occurred, and only light below a threshold wavelength, or above a threshold frequency, was sufficient to initiate the effect, even when the light source was increased in intensity. Considering that the total energy transferred to the surface depends on the intensity of the light, it appeared that there was something else at play, other than the total energy delivered, when electrons were ejected. Furthermore, since light waves transfer energy in a continuous flow, electromagnetism predicts that as the intensity of light is reduced, there should be a delay in the rate at which electrons are emitted, as they take longer to accumulate sufficient energy, yet no such delay was found.

Einstein reasoned that the photoelectric effect can only be understood if we take the light incident on a surface to be composed, not of a continuous flow of waves, but as a collection of discrete wave-packets. As such, each wave-packet—each photon—would be able to cause the ejection of one electron in the surface, and so the intensity of the light—the number of incident photons—would be proportional to the degree of emission—the number of ejected electrons. Furthermore, Einstein's solution explained why there is a threshold in the frequency of light required to initiate the effect. Since each photon contained a minimum amount of energy—a quantum—proportional to the frequency of the light, and since each photon can bump out just one electron, any photons without enough energy to overcome the binding energy of the electron within the surface would not contribute to any effect.

Quite understandably, Einstein's theory disturbed a number of physicists in the early 20th century, many of whom believed that he must simply be mistaken. In particular, Robert Millikan was so unsettled by Einstein's prediction that he spent the next nine years performing experiments on the photoelectric effect that would prove Einstein wrong. However, in 1914 Millikan's experiments had reached a conclusion, and it was revealed that Einstein's theory was remarkably precise. Einstein and Millikan were both awarded the Nobel Prize in physics, 1921 and 1923 respectively, for their contributions in showing that light was describable in terms of particles.

Complementarity

This of course paints a very confusing image regarding the true nature of light. In 1888, it had been empirically proved that light was composed of waves, and in 1914, it had been empirically proved that light was composed of particles. Both of these proofs are positive in their description of the nature of light; light *is* in fact wavelike, and it is in fact particulate. The situation was heightened further in 1923 when Louis de Broglie showed that this seemingly paradoxical description applied not just to light, but to all matter as such.[20] Everything material can be described in terms of waves as well as particles; even the human body is theoretically describable as an arrangement of particles, and as a bunch of waves, for wave–particle duality is a universal feature of the natural world.

Both the wave-view and the particle-view of light and matter are true from one perspective, while false from another, meaning that light and matter cannot be explained classically as either waves *or* particles. Quantum mechanics was initiated by these findings, and it is the central task of quantum theory to describe the mechanisms by which a concrete entity can possess such paradoxical attributes. As Einstein stated:

> It seems as though we must use sometimes the one theory and sometimes the other, while at times we may use either. We are faced with a new kind of difficulty. We have two contradictory pictures of reality; separately neither of them fully explains the phenomena of light, but together they do.[21]

And now, this brings us to the central point of this chapter, which is the presence of complementarity at the heart of physics. In 1927, at the International Physics Congress in Como, the Danish physicist Niels Bohr presented an interpretation of these contradictory findings by way of his principle of complementarity.[22] Bohr had just learned of a forthcoming publication by Werner Heisenberg concerning his recently formulated uncertainty principle. Heisenberg had discovered that the dual nature of quantum objects gives rise to limitations in how accurately we can measure and define the properties of those objects. For example, the more accurately we define the position of a quantum object, the more it looks particulate and the less it looks wavelike, so that we cannot use the object's wavelength to define its momentum. On the other hand, the more accurately we define the momentum of a quantum object, the more it looks wavelike and the less particulate, and since waves are spread out over space, we then cannot define its position.

The essence of Heisenberg's uncertainty principle is that the indeterminacy of either the position or momentum of an object will approach infinity as the other approaches zero. To know both of the properties fully and simultaneously would be like moving each side of a seesaw to the ground, and as we know, if one side goes up, the other must go down. We now understand that the uncertainty principle applies to a wide range of properties of quantum objects—position and momentum, energy and duration, coordination and causation, coherence and entanglement, spin on different axes—and since all objects are describable quantum mechanically, the uncertainty principle is a principle of physical reality in general.

Inspired by Heisenberg, Bohr described how such properties exist in complementary pairs, and that the physical world manifests as an exchange and balancing of these pairs of properties, only one of which can be expressed fully at one time. The properties, as measured through differing experimental arrangements, are classically irreconcilable while also being mutually necessary for an exhaustive description of the respective system. As such, the different pieces of information must be thought of as complementary, and their mutual exclusivity is a result of a limitation in using classical concepts to describe and interpret quantum phenomena.[23] Complementarity is therefore a momentous principle for physics, and arguably the central mass around which the whole of quantum theory orbits. In the words of the American physicist John Wheeler, "Bohr's principle of complementarity is the most revolutionary scientific concept of this century and the heart of his fifty-year search for the full significance of the quantum idea."[24]

The essence of Bohr's principle is that the two participants of complementarity are essentially harmonious descriptions, and in this way, complementarity isolates itself from the Hegelian notion of sublation or synthesis.[25] Complementarity does not distinguish the two poles of a single spectrum, but rather concerns the relationship between two distinct properties, each inversely correlated with the other. Just like a seesaw, there are two directions of linear motion, and while each side of a complementary relation is diminished as it is *replaced* by its opposite, it does not literally *become* its opposite. As such, the duality is necessary, and there is no room for a synthesis.

Bohr's principle has more in common with the thinking of Kant, and even Hume, as opposed to Hegel. In his initial presentation of complementarity, he described how the recent findings in physics called into question our understanding of reality as a mind-independent object. It is not merely that we are unable to measure both properties of quanta at once, but that only one of these properties can be said to exist at one time. There is a limitation in how reality reveals itself to our means of observation; some systems of measurement favour the wave-view, and some favour the particle-view, but as soon as we change the experiment in some way, the world may begin to behave differently, and the determinacy of some properties may be exchanged for others.

The means of observation is just as much of an important element of the empiric process as is its outcome, and Bohr was particularly concerned with the relation between the structures of human cognition and the objectivity of our observations. This is a distinctly Kantian consideration that Bohr saw to apply much more generally than solely to atomic physics. He writes:

I hope...that the idea of complementarity is suited to

> characterise the situation, which bears a deep-going analogy to the general difficulty in the formation of human ideas, inherent in the distinction between subject and object.[26]

Bohr reframes Kant's claim of the equal rationality of the determinism of the objective world and the existence of free will in his application of the principle to biology.[27] In a lecture he gave in 1932, Bohr posited that, in seeking a full understanding of the human organism, the reductionistic approach of molecular biology, which treats lifeforms as collections of mechanisms, and a more holistic functional approach to biology, which treats organisms as whole systems working towards particular goals, are basically incompatible while both being necessary for a complete description of a living being.[28] For Kant, the contradiction implied a schism between the phenomenal and the noumenal, but Bohr revealed it to be a determinable aspect of the former alone.

The uncertainty relations are consequences of wave–particle duality, but Bohr saw all dualities to be effects of a deeper limitation in our ability to describe phenomena. This limitation makes complementarity a general epistemological principle rather than a consequence of quantum mechanics, as Bohr explains:

> For describing our mental activity, we require, on one hand, an objectively given content to be placed in opposition to a perceiving subject, while, on the other hand, as is already implied in such an assertion, no sharp separation between object and subject can be maintained, since the perceiving subject also belongs to our mental content.[29]

It is a characterising feature of the mathematical sciences to remove "all reference to the perceiving subject", and this methodology is limited by the fact that a single objective phenomenon requires "diverse points of view which defy a unique description". Bohr was conscious that, while we are most clearly confronted with this fact in atomic physics, it is a common element of those softer sciences, such as sociology and psychology. Bohr reiterated the point numerous times, and in recounting his many discussions with Einstein on the role of complementarity in science, he writes:

> As is well known, many of the difficulties in psychology originate in the different placing of the separation lines between object and subject in the analysis of various aspects of psychical experience. Actually, words like 'thoughts' and 'sentiments,' equally indispensable to illustrate the variety and scope of conscious life, are used in a similar complementary way as are space-time co-ordination and dynamical conservation laws in atomic physics.[30]

As we shall see, the term 'complementary' had already been used to describe the relation between certain psychological phenomena as early as 1890 by William James, and it is from James' ideas on the subject that we begin in Chapter Four.[31]

Chapter Four

The Opposites in Thinking

Temperament

It has been suggested that Bohr was inspired to the idea of complementarity from his reading of the 19th century psychologist and philosopher William James.[32] However, though Bohr did acknowledge being exposed to James' work in his lifetime, there is no reference to him in his writings. If Bohr was influenced at all in his formulation of the principle, it is likely that this influence was of a more Eastern origin. Bohr's exposure to Daoism—an ancient Chinese philosophy—through his philosophy tutor, is well documented, and it appears that this exposure had a considerable impact on him.[33]

In 1947, Bohr was awarded the Order of the Elephant—the most prestigious honour in the Kingdom of Denmark, and one which is traditionally reserved for royalty. In designing his coat of arms for the award, Bohr chose a *taijitu,* more commonly known as the 'yin–yang symbol', along with the motto "contraria sunt complementa", or 'opposites are complementary'. If anything betrays a deep and long-lasting influence, it is surely this choice, and there is no doubt that the yin–yang philosophy of ancient China conveys the very same reciprocity, interdependence, and mutual wholeness that was central to Bohr's understanding of complementarity in physics.

James' explication of complementarity in psychology is brought to light in a couple of different ways. First, in his 1891 work, *The Principles of Psychology,* James recounts his experiences with neurotic patients who expressed symptoms of psychological compartmentalisation. He notes that these parts of consciousness "co-exist but mutually ignore each other", "share the objects of knowledge between them", and "are *complementary*". James' greatest contribution to the problem of opposites, however, actually concerns the relationship between psychology and philosophy, rather than the constitution of individual consciousness.

James was an incredibly broad-minded and dispassionate thinker—qualities that one would expect to be cultivated via the simultaneous involvement in both philosophy and psychology. But James was particularly pragmatic because he denied that philosophy could ever provide a complete and objective account of truth—that reality is "if not irrational then at least non-rational in its constitution".[34] For James, the nature of one's philosophy is determined by one's attitude, and thus, it is of incredible importance that philosophy should recognise the role of the non-rational elements of subjectivity in the development of its ideas and language. Understandably, while James remains a highly respected figure in the history of ideas, he was vilified by philosophers in his lifetime for he violated the integrity of their conceptions. There is a wonderful passage from James' opening lecture in *Pragmatism*, entitled 'The Present Dilemma in Philosophy', which conveys the message well:

> The history of philosophy is, to a great extent, that of a certain clash of human temperaments...Of whatever temperament a professional philosopher is, he tries, when philosophising, to sink the fact of his temperament...yet his temperament really gives him a stronger bias than any of his more strictly objective premises. It loads the evidence for him one way or the other, making for a more sentimental or a more hard-hearted view of the universe, just as this fact or principle would. He trusts his temperament. Wanting a universe that suits it, he believes in any representation of the universe that does suit it. He feels men of opposite temper to be out of key with the world's character, and in his heart considers them incompetent.[35]

For James, the most powerful premise that a philosopher starts from is a premise originating in their temperament. Since such a premise has no place within the forum, it is largely kept hidden, and this, for James, makes philosophy an insincere endeavour. To be apt for public discourse, our reason must be stripped of all attitude and emotion, yet our recognitions are ineluctably conditioned by our temperament and can never be purely rational. James saw that the general bifurcation of methodology and style, present to various disciplines, reveals a corresponding duality of basic temperaments within the psyche. In politics, it is revealed in the distinction between authoritarians and anarchists; in literature, between purists and realists; in art, romantics and classicists; and finally, in philosophy, between the rationalist—the "devotee of abstract and eternal principles"—and the empiricist—the "lover of facts in all their crude variety".[36] Of course, in discourse we must rely on both fact *and* principle, but a disproportionate reliance on either, James says, leads us into entirely different perspectives on matters of philosophy.

This distinction provides the basis for James' two psychological temperaments: the tender-minded rationalist and the tough-minded empiricist. The former, relying more on principles, tends to be intellectualistic, idealistic, optimistic, religious, free-willist, and dogmatical; the latter, relying more on facts, tends to be sensationalistic, materialistic, pessimistic, atheistic, deterministic, and sceptical. Each of these types sees their way of thinking to be right, and they disparage each other in their basic precepts and values. James sought to express this in the labels he chose for the temperaments: "The tough think of the tender as sentimentalists and soft-heads. The tender feel the tough to be unrefined, callous, or brutal."[37]

The extent to which one's philosophy is conditioned by temperament obviously varies between different philosophical thinkers, and James recognised that some theorists reveal themselves to be more temperamental than others, often those that have very particular and ardent beliefs. In part, James' own philosophy of pragmatism was designed to regain intellectual authority over one's temperament, and to find some middle ground between conflicting opinions. An important message of James' work is that psychology is, if not the starting point of philosophy, at least involved in it; that is because it is ultimately the same mind which is led both to intellectual opinions, and to tastes, talents, and affections.

Acknowledging this, in the remainder of this chapter we shall develop a more comprehensive framework of psychological temperament with special focus on the significance of certain psychological functions for the development of philosophical ideas. Fortunately for us, the hard work here has been done, for the problem of opposites was a central focus for the Swiss founder of analytic psychology, Carl Jung. In Jung's 1921 work, *Psychological Types,* he undertakes a deep and interdisciplinary analysis on what he referred to as the 'type problem' as it pervades psychology, philosophy, and other disciplines in the arts and sciences.

Psychological Functions

Jung's interest in the type problem forms part of his deep and long-lasting preoccupation with the complementarity of opposites in regard to psychic experience. He traced this interest back to his reading of Goethe's *Faust* in his teenage years, and he later adopted the term 'coincidentia oppositorum' to refer to the phenomenon. In English, this means 'coincidence of opposites', in the sense of 'to coincide', and he learned the phrase from a 15th century theologian named Nicholas of Cusa, who Jung referred to as *the* philosopher of the great problem of opposites.[38]

Jung held that a coincidence of opposites was constitutive of the fundamental nature of the self, and that the development of the personality involved both the separation and unification of opposing thoughts and functions. The separation of opposed ideas, feelings, beliefs, and psychological tendencies is mediated by the conscious ego, which, in seeking clarity and consistency for means of survival, identifies itself with one side of a polarity, supressing its antithesis to the unconscious.

Jung explains that discrimination is the essence of the conscious mind, being unable to rationally comprehend contradictions, and that it would be a detriment to survival if it did not learn to facilitate the separation of opposites. This process of separation is compensated by an equal and complementary process of reconciliation, which is the essence of the unconscious. Although we are able to discriminate between opposing ideas in consciousness, those ideas we disregard must become prominent features of the unconscious, and the more these opposites are drawn apart, the more the unconscious seeks out their reunion.

Jung developed his clinical practice in the treatment of neuroses on this principle, for he believed that when the ego attempts to fight off these repressed feelings, tension is created in the psyche, and this tension, if left untreated, is susceptible to developing into illness. The treatment of neuroses, then, according to Jung, lies in the reconciliation of opposing thoughts and feelings, as does the development of the self. Jung wrote about a great number of opposites in relation to the psyche, including general examples like love and hate, action and thought, madness and reason. The most important in relation to the type problem concerns his two basic attitude types, which he coined 'introversion' and 'extraversion' respectively. Jung notes that these terms loosely refer to what James sought to elucidate with his 'tender-minded' and 'tough-minded'.[39]

Note that 'introversion' and 'extraversion' are common terms in modern parlance that are often used to describe one's sociability; in a purely psychological sense, however, they refer more generally to the preferred direction of one's conscious attention. Extraversion is a condition where one's attention is predominantly directed towards objective data, and so the objects of extraversion are generally rather obvious, for these are the objects of our external environment. An extravert of particular sensationalism is largely conditioned by external objects and holds the realism of sensuous reality above all else. As such, the extravert may completely disregard ideals and the assimilation of gathered data as they see sensation simply *as* reality, not looking for any metaphysical understanding beyond that, which they may not consider to be possible at all. Jung refers to this as 'concrete thinking'.[40]

Introversion, on the other hand, is a condition where one's attention is directed more towards subjective views, and so its objects are more elusive for they do not exist concretely in the world. Jung describes that the introvert is focussed on the underlying psychic structure of the mind, which he calls 'the collective unconscious'. This foundational stratum of the psyche consists of inborn archetypes, or primordial universal ideas, that we use to understand our experience of the world. These ideas are recurrent features of our evolutionary history and develop in much the same way that physical organs develop through natural selection. The introvert is conditioned by the contents of the unconscious mind, just as the extravert is conditioned by the object, and they are often unaware of how they have come up with an idea due to the powerful influence of their unconscious.[41] This, Jung calls 'abstract thinking'. He writes:

> The empiricist is always inclined to assume that the abstract thinker shapes the stuff of experience in a quite arbitrary fashion from some colourless, flimsy, inadequate premise, judging the latter's mental processes by his own. But the actual premise, the idea or primordial image, is just as unknown to the abstract thinker as is the theory which the

> empiricist will in due course evolve from experience after so and so many experiments.[42]

Extraversion and introversion are present in all individuals and are both necessary for conscious life, but "the relative predominance of one or the other determines the type" of attitude an individual possesses.[43] As Jung notes, the general predominance that we have in the West is that of a disproportionate valuation of the object in relation to the subject, but the introvert sees the world rather differently, for they value their subjective views over objective facts. This creates a number of challenges for the introvert as, if they are to fit in in their society, they must learn to play by that society's rules, which means joining in with the overvaluation of the objective factor in everyday life. This, in itself, presents some psychological danger for the introvert, for the more they pretend to value the object, the more they betray their own defining principle, which is, themselves, the subject. This is surely the root of a lot of the issues that many introverts face, and it often leads them to view the extravert as their oppressor.

Jung further enumerated two secondary polarities, the 'rational' judging functions, *thinking* and *feeling*, and the 'non-rational' perceiving functions, *intuition* and *sensation*. It is the former pair, given that they are conscious and directed, that will prove important for our consideration of how different individuals become attached to different ideas in philosophy.[44] These functions refer to the means by which one processes information into rational judgements.

Thinking is all about making distinctions, and those who prefer thinking arrive at their judgements based on evidence, information, and reason. They are principally concerned with what makes most sense and tend to detach from situations as to determine the most rational and level-headed course of action. These are people who 'follow their head' and are driven by a dispassionate objectivity; however, because of their reliance on logic they can also appear unempathetic and cold to those who rely more on feeling.

By comparison, feeling has more to do with making connections, and it comes to judgements based on principles, emotions, and their impact on the wider environment. Feeling is principally concerned with the promotion of harmony and so one who favours feeling will tend to empathise with the situation as to better understand the feelings of others. Feeling is dominant in those who 'follow their heart' rather than their head, but this does not at all mean that such individuals are irrational, it merely means that they value harmony over their capacity to be right, placing little value on proving others wrong. They are driven by a passionate subjectivity, but due to their reliance on emotion they can indeed be susceptible to fallaciousness and sensitivity.

The distinction between thinking and feeling is a model example of complementarity in relation to psychology, as it's clear that these functions lead into quite different perspectives on the world, but also that both are necessary for the forming of well-informed judgements. As James recounted: "Facts are good, of course—give us lots of facts. Principles are good—give us plenty of principles."[45] Indeed, any approach to psychology that disregards feeling is bound to be somewhat heartless, while one that disregards thinking will inevitably be imprecise.

So now we have distinguished two dimensions of psychic activity, composed of the attitude-types—extraversion and introversion—and the judging-types—feeling and thinking—both of which express a complementary relation. Since these functions represent two different dimensions of psychic activity, an individual's preference within each is combined to form what Jung referred to as a 'psychological type'. These types can now be represented as a dual-axis model, or a quadrant diagram, as shown on the following page.

	Introversion	Extraversion
Thinking	Introverted Thinking	Extraverted Thinking
Feeling	Introverted Feeling	Extraverted Feeling

The four types are: *extraverted thinking, introverted thinking, introverted feeling,* and *extraverted feeling*. Our eventual goal in the first part of this study is to understand what kind of philosophical theories are *correlated* with each of these psychological types, in terms of the structure of the dual-axis model. I emphasise the word 'correlated' here because it is *not* our intention to attempt typification of the supporters of respective philosophical theories. A goal like this would be empirical, require extensive examination, and there is no guarantee that we would find any meaningful connection between one's psychological type and one's philosophical beliefs. Of course, this was precisely what William James sought to do, and I don't think his contention is unreasonable, but I must be clear that the present study does not depend on there being such a connection.

Moreover, we know from the historical literature that it is entirely possible to hold beliefs from both sides of James' dichotomy. George Berkeley, for example, was both an empiricist and an idealist, and Spinoza was a rationalist with, at least, materialist tendencies. Jung's types are relevant to all aspects of conscious experience, and if we were to study the psychological inclinations of professional philosophers, it would not be too surprising if we were to find a general bias towards the introverted thinking type, purely due to the nature of the profession.

Of course, even the introverted thinker may be led to beliefs that are 'symptomatic' of extraverted feeling if that is where their thinking leads them. I reiterate, therefore, that our aim is to show that certain philosophical positions are correlated with the *use* of respective functions, not that someone who *favours* that function should believe in a particular thing. With this said, we shall proceed to a brief description of the four types, and for this our primary resource is chapter ten of Jung's *Psychological Types*.

Typology

The extraverted thinker is they whose thinking conforms to objective facts, which is to say, whose ideas come from external conditions rather than internal ones.[46] They centre their life on the intellectual conclusions gathered from all around them, and these conclusions rule over all of their perceptions and judgements. That which aligns with them they see as good and proper, anything that does not, they abhor. The industry of peer-review literature is a clear product of this type, where academics interact and build on each other's ideas in the formation of consensuses. This type may be reinforced by a penchant for sensation, serving to further increase their attachment to the object.[47] Such are those who give off the outward appearance of unparalleled reason, but internally their intellect is suffocated by a disregard for ideas and possibilities. They are the apotheosis of what we have referred to previously as the 'objectivist' in regard to philosophy.

The introverted thinker is they whose thinking derives from

internal conditions.[48] They are principally concerned with the development of ideas and theories, rather than empirical practice, and so they are the stereotypical philosopher, mathematician, or theoretical physicist. Due to their self-determining belief system, they are most prone of all the types to controversial opinions, and their introversion may make them appear secretive if these opinions are never revealed. Their privacy is not personal or deceptive, however, for they are simply too preoccupied with the development of their ideology to seek the advantage of others. Nevertheless, their absorption in their own ideas can lead them to be bewildered when others fail to understand them, and this can lead them in turn to conclude that others simply aren't as intelligent as they. This, of course, may or may not be the case, it is rather that others have not thought about the same things so deeply, purely because they are not as interested.

The introverted feeling type are they whose feeling is determined by subjective factors.[49] This is an allusive type because their feeling is directed inwards, rather than outwards at the world, and so they have no desire to impart themselves on others. The internalisation of their feeling can also make them appear cold, but it is rather that they just do not express themselves in a way that would make people think otherwise. The introverted feeler is someone who feels strongly and is conditioned by qualities and principles, like morality, peace, and beauty. Jung explains that, unlike the extroverted feeling type, the introvert's feelings develop in depth rather than in scope, and thus when the introvert does express their feeling function, it is normally in the form of heroic acts of kindness or sacrifice, and much to the surprise of those around them. If this type is accompanied by a strong capacity for intuition, they may become so fully devoted to their subjectivity that they renounce traditional virtues entirely, and their intuition begins to manifest as profound spiritualism or religiosity. Such individuals are at danger of falling victim to the power of objective reality, as all mortals must do. This type epitomises what we have previously referred to as the 'subjectivist' in relation to philosophy.

Finally, the extraverted feeler is they whose feeling is oriented by external factors, and so they seek harmony with objective or socially accepted values, rather than those arrived at personally.[50] Such values are the basis for all things fashionable, as well as for the support and advocation of social and philanthropic movements. In this sense, they value responsibility above all else, striving to provide service, to uphold the good, and to defend against injustice. These are social butterflies who are open with their opinions, but their devotion to rule and conformity *can* lead them to want to influence others. Yet, they are not oppressive, and their affectation comes purely from a place of altruism, which can even lead them to ignore their own personal needs.

Abjectivism & Superjectivism

Now, if we return our attention to the quadrant diagram, we can see that the spectrum between subjectivism and objectivism extends from the bottom left, at the extreme of the introverted feeling quadrant, up to the top right, at the extreme of the extroverted thinking quadrant. The two remaining quadrants then express a second spectrum extending from top left, at the extreme of the introverted thinking quadrant, through to the extreme of the extroverted feeling quadrant at the bottom right. Psychic activity that falls along this line is neutral in regard to subjectivisation and objectivisation respectively, but it is distinguishable in a different sense.

We can make this distinction by understanding that the first spectrum, between subjectivism and objectivism, is really two spectrums, defined by the x and y axes of our diagram. The x-axis, from left to right, then represents the degree of objectivisation present in the psychic process, and the y-axis, from top to bottom, represents the degree of subjectivisation. Accordingly, we can consider that the extreme of the subjective type, at the bottom left corner, combines a maximal favouring of the subjective, and a maximal disregard of the objective, while the extreme of the objective type, at the top right corner, combines a maximal favouring of the objective, and a maximal disregard of the subjective. We can now understand the remaining two quadrants based on their location upon these axes.

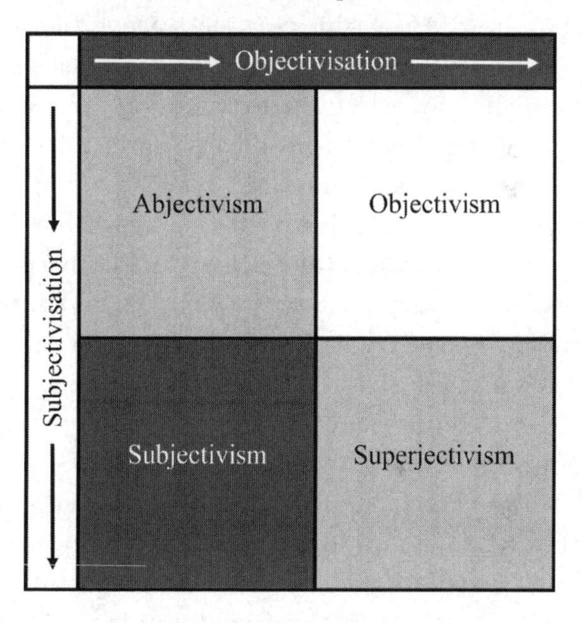

First, the quadrant at the upper left—introverted thinking. This corresponds to the disregard of both subjective and objective factors, and this relates fittingly to the description of introverted thinking provided by Jung. To reiterate, the introverted thinker does not rely explicitly on the ideal objects of subjectivity, nor are they committed to the empirical objects of their environment. Rather, the introverted thinker is concerned with ideas and theories they have derived using logic and reason. They may play on information provided by the object, and they may play on qualities realised in their subjective experience.

However, their ability to manipulate these ideas into their own perspective is the principal determinator in their temperament, and as such, they neither subjectivise nor objectivise their conscious experience.

Opposing the introverted thinking type is that of the extroverted feeler, which expresses the exact converse. This quadrant corresponds to the presence of both subjectivisation *and* objectivisation, also relating fittingly with the description provided by Jung. Again, this type seeks above all else to find harmony between external conditions and internal sensations. They are oriented by objective reality, but only insofar as they can align to it the abstract and universal principles that arise from the unconscious. Extraverted feeling is a particularly intriguing temperament for it is not discouraged by the apparent contradictions between objective facts and subjective principles, instead looking always for their fusion. It is no surprise that this should be the most judged of all the types, for the ambition to cultivate such harmony is bound to seem unreasonable to those more biased or sceptical types that see their perspectives standing in stark contrast to their antitheses.

With this understood, we can now give each of these quadrants a label, and for this we can simply borrow the Latin prefixes that juxtapose the ones we have already. *Sub-* means 'under', and *ob-* means 'against' or 'towards'; so, if we take 'under' and 'towards' to reflect directions of attention, we can use the remaining directions—'above' and 'away from'—to define our remaining types. For the introverted thinking type, who does indeed turn away from any bias towards the subject or object, the prefix *ab-*, which means 'away from', would seem quite fitting. Of course, 'abjection' is already defined, and though we intend to use this term for a slightly different purpose, there are significant similarities with the current usage. The term 'abjective' literally refers to that which is *cast away* from experience, opposing the 'objective', which is *in the way* of it.

Bulgarian–French philosopher Julia Kristeva developed the concept of abjection as the feeling one experiences when confronted with a breakdown in the distinction between what is self—subjective—and what is other—objective.[51] The term is used to denote a feeling of horror felt when faced with the inhumane or repugnant, but here we are using it to describe a perspective where one looks beyond the subject–object relation, potentially at a more fundamental stratum of reality beneath it. It does not therefore refer to the horrific in terms of that which is necessarily frightening, but rather that which disturbs the collective standards which make way for order and harmony. It is therefore also related to the notion of the taboo in society, and this fits accordingly with a psychological type that seeks accord neither with that which is apparent in the subject nor the object, leading more willingly into controversy, and indeed taboo.

The term also contrasts well with our second term, which relates to extroverted feeling. The prefix *super-* means 'above', and so 'superjectivism' refers to that which oversees and encompasses the dualism between subjective and objective experience. Indeed, Kristeva developed her notion of abjection to complement Freud's notion of the *superego,* which refers to a kind of cultural conscience that is learned from external factors such as education and conformity with social norms. There is therefore some similarity between our 'superject' and Freud's 'superego', for superjectivism does indeed involve a desire for social order. Kristeva's relation is expressed succinctly in her aphorism: "To each ego its object, to each superego its abject,"[52] which we could rephrase as 'To each subject its object, to each superject its abject.'

The spectrum between abjectivism and superjectivism does not express any bias towards either the subject or the object, but rather in how fully those factors find expression. The abjectivist neglects the subject–object relation in favour of some other rationality, while the superjectivist seeks to serve it in accordance with the socially accepted good. These four classes of psychological functions will later form the foundations of our investigation into philosophical concepts. However, though Jung developed his notions through the empirical study of patients, psychology still remains a one-sided and subjectivist science. His treatment of neuroses, for example, is an idealistic process, and many in the medical community would argue it ignorant of the decidedly material regulatory process existing within the brain.

As such, it will be necessary to briefly inquire into whether a correlation can be found between Jung's types, and that other approach to human cognition—the objectivist perspective, which forms the basis of biological psychiatry.

Chapter Five

Neural Bifurcation

The Autonomic Nervous System

Over the past century or so, it has been repeatedly suggested that some part of personality is determined by the regulation of one's autonomic nervous system. First, in 1915, the physicians Leo Hess and Hans Eppinger suggested that a disturbance in the balance between sympathetic and parasympathetic activity might be the cause of certain pathological diseases and neuroses.[53] Later, Hans Eysenck suggested that introversion and extraversion are determined by nervous system sensitivity, and that sympathetic dominance was the physiological basis of neuroticism.[54] Jung himself asserted in a letter dated 1939 that: "The unconscious is largely identical with the sympathetic and parasympathetic nervous system, which are the counterparts of the polarity of unconscious contents."[55] More relevantly, in 1974, the British psychologist David Lester proposed that increased sympathetic activity is a major mechanism of extraversion, while parasympathetic predominance is found in the introvert.

The sympathetic and parasympathetic nervous systems together form the autonomic nervous system, which regulates our bodies' unconscious actions. The sympathetic gets the body ready for action by increasing the heart rate and constricting blood vessels, which in turn increases blood flow to the arms, legs, and organs involved in physical activity. The parasympathetic, on the other hand, acts to counter the functions of the former, relaxing the muscles, initiating metabolic processes, and generally promoting a more relaxed, recovery-focused state of being.

What is compelling about the relationship between sympathetic and parasympathetic activity is that each antagonises the other. While both remain active in the background to maintain homeostasis, only one can become fully active at one time, lest our bodies attempt to fulfil conflicting actions, such as simultaneously increasing and decreasing blood flow. We also could not survive without both systems; without the parasympathetic we would rapidly burn out of energy, and without the sympathetic we would enter a state of paralysis. This mutually dependent yet oppositional behaviour is precisely in accordance with the complementarity principle envisioned by Niels Bohr.

In seeking a physiological basis for Jung's ideas, it is not unreasonable that sympathetic and parasympathetic activity could play a role in one's attitude. Indeed, Lester has produced evidence that significant markers of extraversion and introversion are associated with sympathetic and parasympathetic predominance respectively.[56] Moreover, the activity of each of these sides of the autonomic nervous system is regulated by neurotransmitters that shed more light on the psychological differences between these temperaments. The sympathetic is activated principally by the stimulating neurotransmitter dopamine, also the chief modulator for the brain's reward system. The primary neurotransmitter for parasympathetic activity, on the other hand, is acetylcholine, which, conversely, slows down the body, heart rate, and other homeostatic functions.

The interactivity of these two neurotransmitters reveals a genetic component to temperament, as it has been found that one's sensitivity towards the neurotransmitter dopamine strongly influences novelty and risk-seeking behaviours.[57] Several genetic polymorphisms are known to influence this sensitivity, such as the gene that encodes the length of D_4 dopamine receptors, and in a gene that encodes the rate of dopamine metabolism, in turn reducing dopamine activity and upregulating sensitivity. *Reduced* dopamine sensitivity is thought to be a mechanism of extraversion insofar as more stimulating activities and environments must be sought out to stimulate the dopaminergic pathway.

The introvert, on the other hand, may find such activities *over*stimulating and anxiety inducing due to their *increased* sensitivity, leading to a drainage of energy, and subsequent aversion to such environments. For them, the subtler activity of acetylcholine, which is also involved in reward response, may be more enjoyable. Acetylcholine is highly involved in promoting learning, alertness, concentration, and motivation, rewarding us more when we turn inwards. When we contemplate possibilities, visualise, and think philosophically, we are utilising the cholinergic pathway, and these can be deeply rewarding activities for the introvert. The extrovert, conversely, with their focus on the obvious and impressive charge of dopamine, may not even notice when they get a hit of acetylcholine, and this could lead them to devalue contemplative activities as boring, pointless, or unpractical; much better to be out there in the world interacting physically than to sit pondering possibilities.

The Lateralisation of Cognitive Function

The second pair of psychological functions to consider are thinking and feeling, and the most likely physiological correlation for these functions was identified in the 1960s by the neuropsychologist Roger Sperry. Sperry conducted research on split-brain patients—a condition where the bundle of nerves connecting the cerebral hemispheres is severed, leaving each half to function autonomously. The bisection of the brain is a medical procedure that is sometimes used to treat severe cases of epilepsy. With each hemisphere unable to communicate with its partner, Sperry was able to undertake experiments on such patients to study the operations of each hemisphere in isolation. This is possible because each hemisphere of the brain controls the movement and vision of the opposite side of the body, and by presenting words or objects to one side of the visual field, or by closing one of the patient's eyes, a single hemisphere is forced to interpret and process the visual stimuli.

Sperry found that depending on which side objects, images, or words were presented to the participant, different responses were produced, thus revealing that the hemispheres process information in different ways. For example, when patients were presented with two different objects, one to each side of the visual field, and then asked to draw what they saw, participants drew what they saw through their left eye, processed by the right hemisphere. However, when they were simply asked to recall the object they were presented with, participants described what they saw through their right eye—processed by the left hemisphere. Furthermore, when participants were shown a word to their left eye alone they were unable to recall it once it was removed, and, based on a number of experiments of this kind, Sperry concluded that the left hemisphere, but not the right, was capable of recognising and using speech and language, while the right had a better time dealing with images.[58]

Following Sperry's results, for which he was awarded a Nobel Prize in 1981, other researchers set out to replicate and expand on his findings. Studies found that the left hemisphere was more specialised towards analytical and verbal functions, while the right was more specialised towards perceptual and emotive functions.[59] The lateralisation of cognitive functions quickly became a central interest of pop-psychology, and several publications emerged attempting to generalise experimental findings, claiming there to be a clear division of function between the two hemispheres—the 'logical left' and the 'creative right'. In reality, complex functions like logic and creativity are neurologically global phenomena, and it is the finer details of complex functions that are lateralised between the hemispheres.

Regardless of the oversimplifications that are still disseminated today, the two hemispheres of the brain do have consistently diversified functions, which manifest to answer two rather different sets of challenges the evolving species must face. Even the 700-million-year-old sea anemone has an asymmetrically bifurcated neural network, and it is indicative for any inquiry into the nature of the oppositions of philosophy that evolution should divide the nervous system so assiduously. It is less efficient for information to shuttle through the corpus callosum for the hemispheres to function synchronously, and it is remarkable that each is capable of functioning autonomously after their connection has been cut. Indeed, complementarity appears to be hardwired into us by the biological forces of nature.

Complex functions are shared between the hemispheres, reflecting the fact that complementarity expresses similarity through differing modalities, and so lateralisation is a duality in the ways in which a single goal is achieved. Research suggests that the right hemisphere focusses more on a big picture image, while the left hemisphere focusses more on details. That is, the right features a precedence towards processing the global features of stimuli while the left features greater local precedence.[60] We can understand how this might lead to different approaches to matters of philosophy, for the left considers only what is said, and not what is not said. Values and non-verbal cues thus remain elusive to the left hemisphere, and it may resolve moral considerations in terms of their consequences, rather than by any general principles that could underlie them.

It is not an unreasonable suggestion that Jung's thinking and feeling functions are related to left versus right hemispheric processing, even if these labels prove to be more of a convenience than a technicality. Jung reported that the thinker bases their judgements on an intellectual formula, proclaiming that which agrees with it to be good, and that which disagrees with it to be immoral or wrong. All activities—art, science, maths, and literature—rely on both functions, but the roles that each play provide a different perspective on the task. Language is highly localised in the left-hemisphere, as is critical and comparative analysis, while social information processing, spatiotemporal awareness, and the recognition of novelty is thought to be predominant in the right.[61]

In the 1980s, Jungian psychologist Katherine Benziger, in collaboration with the neuroscientist Karl Pribram, inquired into the specialisation of the two hemispheres in an effort to determine where respective functions were located. Benziger's conclusion, which was reached via a combination of PET scan analysis and psychometric assessment data, was indeed that Jung's thinking and feeling functions are localised in the left and right hemispheres respectively.[62] Specifically, she presents that the frontal left, particularly the prefrontal cortex, was specialised towards the 'objectivistic' thinking function, while the basal right, including the temporal, occipital, and parietal lobes, was specialised towards the 'subjectivistic' feeling function. Jung's non-rational functions are also related to the remaining lobes, with the sensing function located in the basal left, and the intuiting function located in the frontal right, resulting in a distinction between the thinking–sensing left hemisphere, and the feeling–intuiting right hemisphere.

Benziger further posits that the habituation and development of a function which is not one's natural inclination leads to excessive energy consumption in the brain, and consequential stress, anxiety, and exhaustion—Jung referred to this as the 'falsification of type'. Benziger's is an interesting approach to the type problem since it asserts a crucial distinction between one's functional competencies, which are cultivated via practice, education, and social influence, and one's natural gifts and talents. In many individuals, the two will be in line, but it is only the former which would be identified via Jungian personality typologies, such as the *Myers–Briggs Type Indicator*, which has become particularly popular in pop-psychology. On this view, such tests more accurately measure functional habituation rather than deep-rooted temperament, which may have become obscured by the former. Benziger's system for identifying and maximising these natural inclinations has become most popular in the corporate sphere, where it seems to have had its greatest influence in Latin America.

Altogether, the brain clearly expresses a complementary lateralisation of cognitive functions, and the fact that it is complementary means that the hemispheres are designed to work in unison. Jung's functions are ultimately generalisations that apply more broadly than the definitions of the words 'thinking' and 'feeling', and so we should not take this to suggest that the left-hemisphere is the thinker, while the right-hemisphere is the feeler. Nevertheless, the clear division of processing styles, and their affinity with Jung's types, is enough to infer that there is at least some connection between the two, and we should use the neurophysiological data to improve our definitions of the types, rather than try and mould it to them.

<table>
<tr><td>Feeling</td><td>Thinking</td><td>Introversion</td><td>Extraversion</td></tr>
<tr><td colspan="2">Judgement</td><td colspan="2">Attention</td></tr>
<tr><td colspan="4">Psychology</td></tr>
<tr><td colspan="4">Perception</td></tr>
<tr><td colspan="4">Neurophysiology</td></tr>
<tr><td colspan="2">Central</td><td colspan="2">Peripheral</td></tr>
<tr><td>Right</td><td>Left</td><td>Parasympathetic</td><td>Sympathetic</td></tr>
</table>

Combining these findings in both the peripheral and central nervous systems, we have now identified physiological correlates for our four psychological functions. Here, we see duality emerging at three different levels, with the lower levels expressing the interdependence of opposites more strongly. At the functional level, the opposites are antagonistic, while the more general relation between attitude and judgement, peripheral and central nervous systems, is conjunctional but independent. The highest-level duality, existing between psychology and physiology, subject and object, mind and matter, expresses a correlation or reflection, as in metaphysical dualism, though we shall not suppose that such dualism is anything more than an appearance.

Part II

On the Structure of Concepts

Chapter Six

The Dialectical Matrix

Continuity & Discreteness

With the quadrant diagram developed in the previous two chapters as our foundation, we are now in a position to inquire into the various concepts and theories we have come up with in philosophy, as well as the relations between them. Our eventual goal is to reveal a uniformity between the structures of philosophical perspectives and psychological temperaments, but only insofar as we can clarify the existence of such a structure, and we shall not posit that the structure is a consequence of psychology. The psychological types, and their physiological correlates, may be taken as expressions of this structure, but it will be prudent to present a version of the model which is not dependent on psychological factors.

We previously saw how the quadrant diagram can be presented as a coordinate system, with *x* and *y* axes representing the degree of 'objectivisation' and 'subjectivisation' respectively. We found a spectrum existing between the subjectivistic *introverted feeling* type and the objectivistic *extroverted thinking* type, and a second neutral spectrum existing between the 'neither' approach of abjective *introverted thinking*, and the 'both' approach of superjective *extraverted feeling*. In order to strip this model of its psychological underpinnings, we will now attempt to understand what might influence a philosophical thinker to become oriented by the subjective or objective factors respectively. Involved in this consideration will be a more general question as to what differentiates the two parts of any complementary relation.

At root, complementarity arises when there is a fundamental limitation in our ability to capture or represent a system with a particular style of definition. However, it is not merely a limitation in our ability to analyse the system, but a limitation in how the system itself can be converted into some other medium. Consider we want to convert an odd number of halves into an integer or give a single solution to a square root; there will always be missing information in the result because there is no single answer to the problem. Importantly, however, there are not three answers, but precisely two, each capturing a significant but incomplete part of the solution. Complementarity does not entail a duality within the underlying system, for the underlying system may be unified and merely incompatible with the mode of human perception and reason. Indeed, the underlying system *may be* innately contradictory, while it is the nature of human perception and reason to represent it with consistency.

The concrete source of the principle of complementarity, provided by Niels Bohr, is given by the contradictory copresence of a wave and particle nature to matter and light. These are continuous and discrete phenomena respectively, as is the distinction between analogue and digital signals, and they reflect opposing modes of expression of a system that is representable as neither alone. Observable quantities are always in the form of discrete values, while motion and change are always continuous. In regard to our capacity to measure and model the external world, continuity and discreteness exhaust our means of description. If reality is in fact describable by neither or by a fusion of these frameworks, then it is incompatible with the manner in which we perceive the object, and with the structure of our language and logic. Currently, the mystery of wave–particle duality remains to be explained, but the complementarity between continuity and discreteness is not merely an artefact of its occurrence in physics. In fact, the recognition of complementarity that set me on this journey was conveyed at a basic level by the abstract relation of continuity and discreteness, and so I had already unearthed an intuitive connection to these concepts long before I learned of their occurrence in physics.

The opposition between continuity and discreteness has played a decisive role in philosophy since its inception. A continuous entity has no internal structure, contains no gaps, and is unbreakable into atomic parts, which is to say, parts that cannot be further decomposed. A discrete series or collection, on the other hand, is decomposable by its nature, being built of individual fragments which *cannot* be broken down into more basic fragments. In the Presocratic period, this distinction was fundamental to the emergence of metaphysics as a topic of study, with the Eleatics claiming reality to be an infinite and immutable continuum, and the atomists claiming it to be composed of discrete parts.

Zeno of Elea formulated several paradoxes regarding the infinite divisibility of motion, distance, and duration in support of the Eleatic view on the impossibility of change, and these paradoxes are grounded in the antinomy of continuity and discreteness. Any finite movement in a continuous space would imply an infinitude of sub-movements, thus consuming an equally infinite duration of time, making any movement impossible. Democritus of the atomist school sought to avoid these paradoxes by positing that matter is composed of discrete atomic parts, forming the lowest possible level of division, and necessitating a minimum possible movement that any object can make.[63]

We then have the development of the calculus in the 17th century by Leibniz and Newton, which exists to summate and discretise the infinitesimal elements of continuous change. Calculus is itself comprised of a complementary pair of operations: differentiation, serving to analyse and break things down; and integration, serving to synthesise and add things up. The dialectic between continuity and discreteness is particularly prominent in mathematics, in virtue of its abstractness, underlying for instance Georg Cantor's assertion that the quantity of the real numbers—the continuum—is the smallest number greater than the quantity of the integers—the discrete series.

The continuum hypothesis seems to get right to the heart of the matter, for it alludes to a kind of union of sameness and difference between the continuous and discrete aspects of infinity, which, perhaps, can only be resolved through mutually exclusive, conflicting, yet complementary viewpoints. The hypothesis was once considered to be the most important unsolved problem of mathematics, and it has since been shown that both the hypothesis and its negation are consistent with, yet unprovable within, standard foundations of mathematics. This may well also be a consequence of the mutual interdependence and reciprocity of the continuous and discrete respectively.

When it comes to characterising the nature of complementarity, then, it seems natural to appeal to continuity and discreteness as modalities through which such reciprocity finds expression. When it comes to characterising our beliefs in philosophy, however, whether our theories express these properties is seldom a matter of concern. We do not argue, for example, that empiricism is favourable to rationalism because its elements are discrete, nor that deontology is preferable to consequentialism because its principles are continuous. Continuity and discreteness are not what I would call *belief-satisfying properties*, which is to say, we are not led to our beliefs purely by their expression of them.

Immutability & Determinateness

One thing that is quite clear from the various examples of complementarity, and theoretical dualities more generally, is that there is a consistent theme between their participants, as expressed in William James' distinction between the 'tough' and the 'tender'. One aspect is persistently more concrete, decomposable, measurable, overt, obvious, and orderly among other similar attributes. This is the particle in physics, digital signals, empiricism, properties, facts, and objectivity. The other aspect of a complementary relationship is consistently more abstract, holistic, intangible, covert, obscure, and chaotic. This is the wave in physics, analogue signals, rationalism, functions, principles, and subjectivity.

The distinction between concreteness and abstractness is particularly apparent within complementary relations, and within it we can begin to understand why individuals might fall on either side of the divide on matters of philosophy—between subjectivism and objectivism. Many of the entities we usually consider to be continuous, such as space, time, and electromagnetic waves, are not tangible things, and we cannot measure them in any classical way. We can measure distance and duration arbitrarily, but we cannot take a piece of time and have it interact with our instruments of observation.

As we saw in our discussion of the photoelectric effect, light can only transfer energy in discrete quanta, and this quantisation is a direct consequence of the measurement process. Any interaction of a quantum object with some instrument of observation causes that object to appear as a discrete entity. We can measure some artifacts of waves, such as interference patterns or diffraction gratings, but any direct, energetic interaction with light or other quanta occurs between two discrete systems. The capacity of being measured is a qualifying mark of concreteness, as it is this same capacity that allows particles and other concrete forms to impart change on others.

The reality of the objective world lies in this capacity to be measured, and the absoluteness of the empiric process, which is the foundation of science, gives power to objectivist perspectives in philosophy. It is only by the fact that concrete objects interact physically and in accordance with the laws of force and conservation that we are able to gather information about their condition. Accordingly, therefore, the basic belief-satisfying property related to discrete and concrete entities is the property of *determinateness*—the capacity of being determined, limited, or otherwise clearly defined.

In defiance of the cold, hard, solidity of corporeal reality, there is another world of philosophical thought where the vagueness of abstracta is more highly valued. Reason need not have a physical referent, qualities and forms may be conceived prior to their instances, and principles can be posited as the basis of moral action. Abstract entities are favoured not because they are determinate, but because they are consistent elements of our experience. Qualities like *redness*, *virtue*, or *beauty* share in common with the continuous their imperviousness to change, this permanence and invariability being the only way we can identify the relation of their instances. It is this sentiment that led Plato to the belief that ultimate reality is constituted of ideal forms, and that sensuous objects are mere shadows of these abstractions.

Qualities have an unlimited number of instances, and though these instances may express the quality to varying degrees, no variance can alter the essential quality of our perceptions. The redness of a rose may differ from the redness of a bird, yet still we can recognise that they share their redness between them, and Plato believed that these instances are imitations of the perfect Form of Redness. In a similar way, there exist in the world no perfect spheres nor forms of any sort, yet we understand very well what it means to be spherical. Plato contends that such ideals are graspable in spite of their indeterminateness because they are immutable wholes existing in a mind accessible stratum of reality.

Qualities are not determinate elements of our experience, yet we utilise them with full awareness of what our terms mean and have no difficulty in positing them as real elements of the world. Regardless of whether qualities are fundamental, or merely conventions of language, the only reason we can talk about them meaningfully is that they are in essence unchanging. And so, the belief-satisfying property existing to complement determinateness is the property of *immutability*—the incapacity for change.

Immutability and determinateness fit the roles of belief-satisfying properties because the entirety of experience can be described by one or the other, such that all immutable forms are indeterminate, and all determinate phenomena are mutable. The division between certain subjectivist and objectivist approaches to philosophical problems lies on this distinction, for abstracta are intelligible insofar as they do not change, and concreta are perceptible insofar as they can be measured. There is a way of looking at both of these properties and finding in them value and credibility as the basis of our beliefs.

We are also able to talk about a pair of properties that work in the opposing direction, to disrupt our belief in certain ideas; these are the negated properties of *indeterminateness* and *mutability* respectively. Material objects are mutable, and this capacity for change is precisely what allows us to make measurements of their properties, as energy is exchanged between the system and the instrument of observation.

Mutability or flux was the ground of Heraclitus' philosophy—"all things are on the move and nothing remains"[64]—and it was in response to Heraclitus that Parmenides of Elea came to doubt the reality of concrete forms purely *because* they appear to change. Even if we do not share the view, it is easy to see how someone who senses more reality in the unchanging might neglect the credibility of objectivist or positivist approaches to philosophy, just as it is easy to see how the empiricist might be led to nominalism—the disbelief in abstract objects—purely because they cannot be determined.

We therefore have a set of four properties: two belief-satisfying properties—immutability and determinateness—and two belief-disrupting properties—mutability and indeterminateness. Over the course of the following chapters, we shall see how the degree to which we are influenced by these properties is related to an ability to find value and validity in subjectivist or objectivist viewpoints respectively. Determinateness influences us towards the objective, and immutability towards the subjective; indeterminateness influences us away from the subjective, and mutability away from the objective.

Each pair of properties consists of an active, positive property, asserting some capacity of the object, and a passive, negative property, denying that same capacity. The positive and active properties are *determinateness,* which asserts the capacity for being determined, and *mutability,* which asserts the capacity for change. Conversely, the negative and passive properties are *indeterminateness,* denying the capacity for being determined, and *immutability,* denying the capacity for change. The opposition between immutability and determinateness then expresses the related oppositions of the passive and the active, the negative and the positive.

The properties can also be related to Jung's psychological functions as follows. Extraversion is a determinate function for its objects are the determinable entities of our external environment; introversion is an indeterminate function for its objects are internal to our experience; thinking is a mutable function for it is concerned with quantities, facts, and compounds of particular ideas; and feeling is an immutable function for it is concerned with universal qualities and principles of experience. Accordingly, the four properties combine in the same manner as the psychological functions to give four property-pairs that are correlated with the types.

	Indeterminate	Determinate
Mutable	Mutable Indeterminate Abjective	Mutable Determinate Objective
Immutable	Immutable Indeterminate Subjective	Immutable Determinate Superjective

In this model, the objectivist is more attracted to the determinate than they are towards the immutable, and less repelled from the mutable than they are from the indeterminate. The subjectivist is more attracted to the immutable than they are to the determinate, and less repelled from the indeterminate than they are from the mutable. The abjectivist is less repelled from both indeterminateness and mutability than they are attracted to their negations, and this is due to the fact that they are even more repelled from taking a position of bias, and most repelled of all from the contradiction that arises from accepting both as equal. Finally, the superjectivist is more attracted to the immutable *and* the determinate than they are repelled from their negations, for they attempt to see these properties not as solely contradictory, but as complementary and compatible.

The Dialectical Matrix

In this second part of the present work, we will learn how this quadripartite structure gives rise to philosophical theories throughout all of the topics of discourse. In this we will find the concepts of immutability and determinateness very useful, for they allow us to identify this structure without any reference to psychology. Temperament should not be taken as a predictor of one's beliefs without evidence, nor should our beliefs be predictive of our psychology. Our aim is to study the correlations between the *structure* of the possible perspectives on a notion, and the general structure of psychic functioning, *without* any assumption of a causality between them. Of course, it is intuitive that in an ideal environment, devoid of any influence regarding the ascription of belief, there should also be a correlation between the temperaments and perspectives of individuals. It is simply not of interest to the present study to claim this.

What we *will* be able to do, providing the model is effective, is evaluate the ideological consistency of a given person's beliefs. As we will use the model the determine the structure of ideas in different areas of philosophy, there will be ideas in different areas of philosophy that relate to the same part of the structure: subjectivism, objectivism, etcetera. If an individual appeals to immutability in regard to both their metaphysical and meta-ethical beliefs, then we may say that these beliefs are consistent in regard to the structure. If, however, a staunch empiricist, who does not believe in anything that cannot be measured, also professes a belief in the objective reality of beauty, we may find that these positions appeal to different belief-satisfying properties and are grounded in opposing parts of the structure. Considerations such as these will ultimately be part of a novel subject of study I refer to as 'syntheorology'.

The uncovering of a structure that informs the ideas we can come up with in philosophy will provide a useful tool for understanding the nature of belief and opinion. The subject matter of syntheorology would thus be the architecture of this structure, referred to from this point as the 'dialectical matrix'. As we shall see, quadrants of the structure generalise categories of perspectives on topics and concepts; specific theories and definitions of concepts can be thought of as points of resonance across the structure; and the four corners of the structure reflect the extremes of each category. Accordingly, syntheorology will be a scientific discipline rather than philosophic, for it seeks only to describe the landscape of opinion, and makes no claims to philosophical truth, nor of the nature of the structure it describes. We will nevertheless find that syntheorology is the tool that will allow us to resolve the problems of philosophy, though on this it is still too early to comment.

What we *can* say is that the dialectical matrix will reveal the structure of our means for reason—the psyche—to be correspondent with the structure of the outcomes of our reason—our theories—demonstrating that both are conditioned by a common factor. This may appear to be an obvious fact, but the monoletheic approach to philosophy can hope for knowledge only insofar as our reason is free, and that we are not constrained to ignorance by the mechanisms of our thought. This in itself would imply, though by no means would it prove, that *if* philosophical truth is graspable by the human mind, *then* it is not monoletheic.

By 'philosophical truth', we are here referring to what Kant would regard as the truth of noumena, and so Kant's position would be that such truth is *not* graspable because *it is* monoletheic. On the other hand, philosophical truth could be graspable because *it is not* monoletheic, and in such a case we would not have to posit the existence of a separate noumenal world, for reality would be compatible, even correspondent, with perception. That is to say, the dialectical matrix could be the structure of reality as well as that of perception and the means and outcomes of our reason.

The latter would be more in line with the thought of Georg Hegel, whereby the rational is actual, and the actual is rational, rebutting the principle of monoletheism. If our concepts are conditioned by the same factor that governs the constitution of the world, then the world shall be contradictory insofar as our reason involves contradiction—complementary insofar as it involves complementarity. The problem produced for monoletheism is not that we can know it to be false, but that we cannot know it to be true, meaning that it has no place in any science of philosophy.

In the following four chapters, then, we shall examine the major concept involved in each of four areas of philosophy, in an effort to ascertain whether these subjects conform to the dialectical matrix. These subjects will be *epistemology*, the study of knowledge and how we might gain knowledge; *ontology*, the study of being and the kind of things that exist; *axiology*, the study of the nature of ethical and aesthetical value; and *political theory*, particularly as it concerns the nature of right and the systems that seek to uphold it. As we shall see, each of these disciplines, though seemingly independent and disconnected, all flourish through the same basic structure, not only expressing a clear division between the subjective and objective viewpoints, but also between that which we have labelled as the abjective and the superjective.

Chapter Seven

On Knowledge

Epistemological Dichotomies

Epistemology is the study of knowledge and is in a large sense the foundation of the whole of philosophy. All of the things we believe to be true are dependent on our ability to gain knowledge of those things, and so epistemology aims to understand the processes involved in gaining knowledge, and how we can provide justification for such knowledge. As we consider what it is about the relation between the psyche and reality, which allows the former to grasp the latter, epistemology presents us with an opportunity for formulating a direct connection between psychology and philosophy.

There are many problems that philosophers tackle in epistemology, but throughout them all is a basic system of distinctions. This system has been conspicuous since the time of the ancient Greeks but was most clearly characterised in the work of Immanuel Kant, particularly in his *Critique of Pure Reason* of 1781.[65] Kant surmised that propositions regarding knowledge can be defined in terms of two dimensions, each formed of a dichotomy. We have an epistemic dichotomy distinguishing the means by which a subject is perceived, and a semantic dichotomy distinguishing the means by which a predicate is justified.

The epistemic dichotomy concerns the experiential process involved in acquiring information, differentiating propositions that are knowable independently of experience—from reason alone—and propositions that are knowable only from experience of our external environment. These propositions are named *a priori* and *a posteriori* respectively, which are Latin terms meaning 'from before' and 'from after' experience.

An example of an *a priori* proposition is the tautology 'all widows are women', since we do not need to rely on experience to know that this statement is true. That is, we do not need to go out into the world and find every single widow to check that they are female; the proposition is justified by pure reason, as all a priori statements are. An example of an *a posteriori* proposition, on the other hand, is the proposition 'all widows are five feet tall', since the truth or untruth of this statement is not obvious by any amount of logic or reason. To know whether the statement is true or not we need to experience it empirically in the world, if not by measuring the height of all widows, at least by observing there are widows of differing heights. Accordingly, apriority and aposteriority are mutually exclusive, all propositions being describable by precisely one or the other.

The semantic dichotomy is concerned with what it is about a knowledge claim that qualifies it as true, distinguishing propositions that are justified by understanding the meanings of the words involved—*analytic*—from propositions that are justifiable only by how those meanings relate to the world—*synthetic*. That is, a proposition is analytic if the concept embodied by the predicate of the proposition—that which is affirmed—is contained within the definition of the concept embodied by the subject—what the proposition is about. Since analytic and synthetic statements are mutually exclusive, we can consider a proposition to be synthetic simply if it is not analytic.

To use our previous examples, the tautology 'all widows are women' is analytic because the predicate concept 'woman' is contained within the subject concept 'widow', and thus we only need to know the meanings of the words 'widow' and 'woman' to know that this statement is true. If we were to replace the word 'widow' with the definition 'woman whose husband has died', the resultant statement 'all women whose husbands have died are women' is quite evidently true by definition. Conversely, the proposition 'all widows are five feet tall' is synthetic because the predicate concept 'five feet tall' is *not* contained within the subject concept 'widow', and therefore we need more information in order to know if it is true.

Now, you may be able to see already that the two dichotomies of knowledge conveyed by Kant appear to be epistemological expressions of the dialectical matrix. Firstly, the epistemic dichotomy, between a priori and a posteriori, is expressive of the distinction between indeterminateness and determinateness respectively, for it refers to whether an object of knowledge is determinable through experience of the world. That is, the a priori is indeterminate because it refers to things that are knowable independently of external determination, and the a posteriori is determinate for the converse. Furthermore, the semantic dichotomy, between analytic and synthetic, is expressive of the distinction between immutability and mutability, for it refers to whether a predicate concept is an immutable property of its subject. That is, the analytic is immutable for it concerns that

	A priori	A posteriori
Synthetic	Synthetic A priori Intuitive	Synthetic A posteriori Empirical
Analytic	Analytic A priori Logical	Analytic A posteriori

which is necessary to the definition of a concept, while the synthetic is mutable for it concerns that which is contingent to the concept.

These two dichotomies can be represented in the manner of the dialectical matrix, resulting in four classes of knowledge: *synthetic a priori, synthetic a posteriori, analytic a priori,* and *analytic a posteriori,* as seen on the previous page.

Again, we can find a spectrum extending from the analytic a priori to the synthetic a posteriori, expressing the polarity between subjectivist and objectivist characterisations of knowledge. We shall now briefly discuss the epistemological notions and perspectives that emerge from each of these four quadrants.

Analytic A Priori

We begin with the analytic a priori, which refers to statements that are justified independently of experience—a priori—and by virtue of the meanings of the words they contain—analytic. Thus, the statement 'all widows are women' is analytic a priori, since its justification requires neither empirical evidence nor information other than the meanings of the words 'widow' and 'woman'. Analytic a priori truths often concern trivial matters that are true in all possible worlds, such as the fact that all squares have four sides or that all runners are moving, and these can never cover truths that depend on how their subjects are expressed. As such, the analytic a priori concerns matters of pure logic, though some logicians, namely Gottlob Frege and Bertrand Russell, have hoped to show that all of mathematics can be reduced to the analytic a priori, removing it entirely from the domain of intuition.

In regard to eliminative belief, due to the tautological nature of analytic truths, it would be a rather restrictive view to hold that knowledge *only* comes through the analytic a priori. Indeed, if I can know anything about the physical world, then I know something that is not analytic a priori. However, the sceptical subjectivist might want to claim that everything one knows about the external world is conditioned by one's own mental states and that propositions about the world can never express genuine truth nor knowledge. René Descartes was sympathetic to this kind of view, for he claimed that the only thing we *can* certainly know is that we ourselves exist.[66] His foundational tenet, 'cogito ergo sum', or 'I think therefore I am', could be interpreted as analytic a priori if we take the concept 'to be' as contained within the concept 'thinking', in the sense that existing is prerequisite of any action.

The belief that all knowledge is analytic a priori expresses an extreme subjectivist approach to epistemology and would be an analytic form of rationalism. Gottfried Leibniz had tendencies towards this view in his ambition of establishing the entirety of human knowledge in a universal language of thought. Descartes' and Leibniz's assertions of the existence of innate truths within the psyche is a form of *psychologism*, whereby metaphysical or logical truths are grounded in psychological facts; and Carl Jung used the term 'ideologism' to refer to the opposite of empiricism, 'rationalism' being for him too general. Neither of these terms are explicit about analyticity, however, so we shall use the term 'analytic rationalism' to refer to a commitment to the analytic a priori. We could also use the term 'logicism', though this is usually associated with Frege and Russell's program of showing that mathematics specifically is analytic.

Synthetic A Posteriori

Directly opposed to the analytic a priori, synthetic a posteriori statements are those which are justified through experiencing the world—a posteriori—and by how the terms they contain are related to external information—synthetic. The statement 'all widows are five feet tall' is synthetic a posteriori because its justification requires empirical evidence. As such, most of what we know about the objective world is synthetic a posteriori in nature, and the belief that knowledge comes to us only or primarily via the synthetic a posteriori characterises the objectivist perspective on epistemology—*empiricism*.

John Locke, who was one of the founders of the empirical method, believed that the human mind is a blank slate containing no innate ideas.[67] From birth our minds are marked by impressions of the senses, Locke thought, with these impressions being our only accessible source of knowledge. Our instinctual ability for things like language is often offered as a challenge to this view, but at a minimum the empiricist believes that sensory information is the principal repository for the acquisition of knowledge, and that anything gathered independently of the senses is unworthy of consideration.

One limitation to this view is that it makes it hard for the empiricist to justify the validity of mathematical claims, which certainly seem to be true, but are not derived from experience of the world. This was also a problem for the logical positivists, since, on Kant's determination, mathematical propositions are synthetic, which for the positivists would make them meaningless. The logicist solution was to reduce mathematics to the analytic a priori, yet the empiricist rejects the a priori altogether, and so this solution does not offer them any recourse.

One option for the empiricist is to affirm that mathematical truths are in fact empirical, and that we gain mathematical insights through collecting empirical data. This was the perspective of John Stuart Mill, and it restricted Mill from holding any mathematical proposition, so basic as 1+1=2, as being graspable through reason alone.[68] Only when we experience two singular objects being united can we know that together they make two. For the mathematical empiricist, therefore, mathematical truths are contingent, and may not obtain in other possible worlds.

The view that all mathematical truths must be experienced to be known to be true is nevertheless difficult to maintain for its more abstract components, such as those that describe complex physical theories. Some philosophers, however, such as Willard Quine and Hilary Putnam, have argued that since mathematics is indispensable for the portrayal of such theories, we must assume the mathematical objects we employ really do exist in nature and are quasi-empirical.[69] The same argument applies to other non-empirical but supposedly physical entities, such as quarks, being indispensable for our descriptions of matter.

However, physical theories are just that—*theories*—and may one day be proven wrong. The fact that abstracta are indispensable for a theory that is likely true does not allow us to conclude that such abstracta exist. Moreover, how 'being required to explain what is observed' is epistemologically similar to 'being almost observable' is unclear, and even if we accept mathematics as *quasi*-empirical, this does not make it empirical simpliciter. While Mill's view is seemingly more radical, it is also most loyal to the empiricist program, and asks us to be conscious of what really is empirical, and what is mathematical extrapolation. Of course, non-eliminative empiricists and positivists alike will accept that we have knowledge of analytic truths, and even the eliminative empiricist can claim that mathematics in the abstract does not express genuine facts.

Synthetic A Priori

Synthetic a priori propositions are those that are justified independently of sensory experience—a priori—and by how the meanings of the words they contain relate to external information—synthetic. If analytic statements tell us nothing genuinely novel about the concepts they describe, and if a posteriori statements convey sense data about the empirical world, then what can account for facts about the world that are not produced by the empirical method? Kant's view was that the possibility of gaining such knowledge is found in the synthetic a priori, and the central task of his *Critique of Pure Reason* was to justify this capability.

Consider the idea that the laws of nature must exhibit the same force across any area of space. Clearly, we cannot measure these forces in every location across the universe, but we can conceive of situations, based on what we know empirically, where it would be absurd for them to not be. Accordingly, if the invariance of physical law is true, then it can only be known a priori, and since 'translation invariance' is not contained in the definition of 'physical law', the claim must also be synthetic.

In *Metaphysical Foundations*, Kant attempts to give an a priori derivation of Newtonian mechanics. As he states, the quantity of matter in existence must always remain constant for no substance is created nor destroyed; and action and reaction must always remain equal for all changes in matter are changes in motion, and the two are kinetically reciprocal. Since the concept of matter does not contain the concept of permanence, we must synthesise two distinct notions in order to derive their relation. As such, while the majority of our scientific understanding is indeed a posteriori, Kant saw that there was another part of science—the "pure part of natural science"—which he recognised as synthetic a priori.[70]

Kant counted on the possibility of gaining metaphysical knowledge of the phenomenal world for he saw certain ideas to reflect necessary conditions for our experience of that world. Hume saw all judgements to be of only two kinds—a posteriori 'matters of fact' and analytic 'relations of ideas'—and since metaphysical necessities can be ascertained via neither, he could not see that such knowledge was possible. By cleaving necessity in two, Kant was able to avoid Hume's fork and found metaphysics upon intuition. All a priori truths are true by necessity, but the synthetic a priori is only necessary insofar as phenomena reflect the a priori intuitions of the mind—intuitions which would not be found in a world devoid of human experience.[71]

Kant also believed that mathematics and geometry are synthetic a priori; for instance, 'the sum of the interior angles of a triangle is a straight line' is synthetic, for the concept 'straight line' is not contained in the concept 'triangle'. Likewise, the sums of numbers are not contained in the definitions of the numbers to be added. Kant of course was unaware that non-Euclidean geometries would be discovered in the years to follow, nor that the logicist movement would seek to ground mathematics in the analytic. Nevertheless, modern foundations of arithmetic do not follow the logicist program and are grounded in synthetic axioms.

The synthetic a priori characterises neither the subjectivist nor objectivist approaches to knowledge, for it concerns neither the determinate empirical objects of our external environment, nor the immutable logical relations of ideas. Instead, it relies on a faculty that is non-inferential, meaning that we cannot infer synthetic a priori truths from other knowledge. The emphasis on intuition is important, for it is the intuitional relation between the subjective and objective, which allows for the identification of those philosophical meta-principles that provide a foundation for science. We can therefore use the term 'intuitionism' to distinguish a commitment to the synthetic a priori, though this term is mainly associated with a position in the philosophy of mathematics, so it may be clearer to use the term 'synthetic rationalism'.

Analytic A Posteriori

The final possible class of knowledge is the analytic a posteriori, which expresses a synthesis of both the subjectivist and objectivist perspectives, characterised by immutability and determinateness respectively. Since these two properties, being complementary, are apparently mutually exclusive, we would not expect that the analytic a posteriori could represent a genuine form of knowledge. This conclusion aligns with the analysis of the analytic a posteriori given to us by Kant, for an analytic a posteriori proposition would be one that must be justified by experience, but which is also self-evident and made true by virtue of the meanings of the words it contains. As per Kant's formulation of the categories of justification, analytic judgements confer necessity, and necessity is a priori since it can be reached prior to experience. It therefore follows that there is no such thing as an analytic a posteriori truth.

Considered on the backdrop of complementarity, to say that the analytic a posteriori is both immutable and determinate is like saying that reality is simultaneously continuous and discrete, or that matter is simultaneously a wave and a particle. The properties explicated by the proposition antagonise each other, and therefore there remains the possibility of understanding the analytic a posteriori from a complementary perspective, rather than a contradictory one—a possibility we shall explore in later chapters and so will withhold judgement on at this time.

And so, this ends our first inquiry into the epistemic categories and classes of knowledge. We can now conclude that perspectives on knowledge can be aptly described by the dialectical matrix, and that adherence to certain epistemic categories is associated with properties governing belief. The value we are capable of finding in certain ideas is proposed to be related to the exercise and influence of respective functions of reason. Before we move on to our next inquiry, a quick comment about the relation between neuropsychological functions and reasoning procedures will be prudent.

It is sometimes suggested that the left hemisphere is more analytical in its processes, while the right is more synthetical. This would imply that the thinking function should be correlated with the analytic on the dialectical matrix, while the feeling function should be correlated with the synthetic, opposing our model. However, the local precedence of left hemispheric processing implies a generalising or bottom-up approach to reasoning that relies on the synthesis of individual facts. Conversely, the global precedence of right hemispheric processing implies a reductive, top-down approach to reasoning that relies on the analysis of parts within a whole. Indeed, there is significant evidence to support the hypothesis that inductive reasoning, which is synthetical, largely activates the frontal left hemisphere, while deductive reasoning, which is analytical, largely activates the basal right.[72] Not only does this conform to the classification of psychological, neurological, and epistemological categories within the dialectical matrix, but it also supports Katherine Benziger's theory on the physiological location of Jung's functions.

Chapter Eight

On Being

Ontological Dichotomies

Ontology and epistemology are intimately connected disciplines, for the possibility of gaining knowledge of an object is bound by the kind of objects that exist. The dichotomy at the heart of ontology is not only the most immediate duality present to our experience, but also constitutes one of the oldest dialectics in the history of philosophy. This dialectic concerns the nature of *substance*, where substance is an ontologically basic property bearer from which perceptible forms are constructed, and that which remains unchanged as properties arise and alter.

The early Greek philosophers were predominantly materialists, proclaiming that nature is at root a material construct. The naturalism of the Presocratics reached its peak with the ancient atomist school, which purported that the natural world, all the states of matter, and even human consciousness, are made of discrete, indivisible atoms of concrete substance.

Materialism was not a universal view among the Presocratics, for Pythagoras and Parmenides believed the true nature of reality to lay beyond the appearance of corporeality. For Pythagoras, this reality consisted in numbers and mathematical principles, while for Parmenides it was an immutable, continuous, and unitary 'Being'. The demarcation of appearance from reality was highly influential in the thought of Plato, who provides the earliest explication of the fundamental ontic dichotomy, and also the clearest antithesis to Presocratic materialism. For Plato, it is the ideal forms and qualities of objects that are ontically basic—qualities like beauty or squareness—while sensations are imperfect reflections of the Forms in which they participate.

Plato's distinction between 'Forms' and 'sensibles' evolved

from the Enlightenment period, as seen through the works of Locke, Kant, Hegel and others, into the modern distinction between *abstract* and *concrete* objects around the start of the 20th century. Gottlob Frege argued that numbers are not merely ideas in people's minds but are nevertheless immaterial. Certain abstractions must therefore be distinguished from those that exist only in virtue of the mind, and there must therefore be a realm apart from the subject and object, which is explored within subjects like mathematics.[73]

Today, abstracta are typically demarcated from concreta on account of their non-physicality, rather than non-sensibility, given that some entities described by physical science are non-sensible regardless. An abstract object is therefore one with no location in space or time and is incapable of imparting change on the objective world.[74] Concreta conversely are spatial and causally efficacious. The abstract–concrete dichotomy is related to the dimension of determinateness in the dialectical matrix, for while concreta are objects we can measure and clearly define, abstracta we cannot.

The second dichotomy that we must consider is also expressed in Plato's theory of Forms; this is the distinction between *universals* and *particulars*. Plato's commitment to the Ideas or Forms is founded in a recognition that such Forms are universals that find instantiation in many particular things. As such, universals are qualities that objects have in common, such as *redness, roundness, speed,* or *silence*. How to account for the existence of universals is one of the oldest problems in philosophy, for it is only in reference to universals that we may describe what things are like, and how things sharing in qualities are *alike*.

Opposing Plato, Aristotle denied the Forms an independent existence, claiming instead that universals like 'redness' exist only insofar as there are red things existing in the world. However, if every quality belongs purely to the object in which it is found, then there should be just as many qualities as there are their instances in the world. In other words, qualities are not universals, as Plato supposed, but particulars. As we will see, this line of thought provides the basis for one response to the problem of universals called 'trope theory'.

The distinction between universality and particularity is related to the dimension of *immutability* within the dialectical matrix. Universals, being pure, continuous, and incomposite wholes, are immutable, and just like the analytic necessities of reason, we are able to grasp universals because of their permanence. For Plato, we can only truly know things that do not change—that are one throughout all their expressions—and this is why he believed that the real world consists of universals. Particulars, on the other hand, are incomplete and composite. Objects may express beauty, but no object is purely beautiful, and so particulars always arise in the object in the presence of their opposites. Their composite nature means that they are continuously subject to change, exchanging properties within themselves, arising and ceasing to be.

	Abstract	Concrete
Particular	Abstract Particular Mathematical	Concrete Particular Material
Universal	Abstract Universal Ideal	Concrete Universal

With this we have identified the four basic categories of ontology, which can be represented on the dialectical matrix to generate four classes of object: the *abstract particular*, the *concrete particular*, the *abstract universal*, and the *concrete universal*.

Here, the polarity between subjectively and objectively realised forms of being extends from the abstract universal to the concrete particular, and we can also begin to see some early correlations between the elements of ontology and epistemology respectively. These correlations will be properly explored in Chapter Eleven, but I will mention them briefly as we continue with a description of the four classes of object and the ontological theories that devotion to each engenders.

Abstract Universal

The abstract universal refers to non-physical qualities that are instantiated in many different things, and we can say that objects are similar insofar as they share universals between them. 'Redness', for instance, is both abstract and universal, being expressed in sensible things, but not being sensible itself. The universal is therefore defined by its unitarity and wholeness—that all red things instantiate a single property of redness.

Abstract universals comprise the same quadrant of the dialectical matrix that gives rise to the analytic a priori in epistemology, and the question arises as to whether, and if so how, these concepts are related. One route to finding such a connection lies in the Kantian concept of a 'mark', which refers to the impressions we have of things, allowing their identification.[75] Kant took an Aristotelian approach to the problem of universals, but he also acknowledged that we can cognise the properties of objects as belonging either directly to the object, as synthetic parts, or as part of a universal concept, in which case the connection is logical, analytic, or as he put it, 'discursive'. For example, perceived redness can be predicated as a property of some phenomenon, where it is "immediate and singular", or as an instance of the universal 'redness', where it is "mediate and general".[76] Houston Smit, an expert on Kantian philosophy, writes, "Kant maintains that... we *make* a representation general through certain 'logical acts' of comparison, reflection, and abstraction," which is to say, by comparing phenomena to judge how they are similar.[77]

Another potential connection between the abstract universal and the analytic a priori lies in Plato's theory of how we have knowledge of the Forms, which he offers in the *Meno*. In this dialogue, he has Socrates proclaim that we can only inquire into that which we already know, since to pose the question, for example, 'What is virtue?', we must already have some idea in mind of what it is we seek. Socrates is sceptical about the possibility of one man imparting knowledge on another, and so the sole possibility of having knowledge of virtue is if that knowledge comes from within.[78] This is the basis of Plato's 'theory of recollection', which posits that universal forms exist innately within consciousness, and can be accessed through a process of remembering. Plato's reasoning is of course systematic and must be recognised in the context of his wider metaphysic, but what the theory does do is highlight the possibility that there is some *analytic* process involved in the recognition of universal qualities.

For Plato, knowledge of the Forms is not learned but is rather available from birth. We may not be able to define, speak, teach, or learn what virtue is, but we shall surely recognise it when we see it. In other words, one *has* the concept 'virtue', and even though we cannot produce a rational description of it that does not depend on its instances, we are able to recognise those instances when we meet them in the world. Furthermore, we understand that virtue is not just virtuous things. As Plato argues in the *Phaedo*, our ability to experience and judge the imperfection of virtuous things, means we must have a priori knowledge of virtue already inside us.[79] It may well be a synthetic, intuitive process that allows us to attach our concepts to sensuous experience, but the presence of those concepts in the psyche is itself a logical necessity. Abstract universals are, in a sense, the axioms of the mind, allowing for our qualitative recognition of the world.

The belief that abstract objects exist and are not purely mental constructions is referred to as Platonic realism, and the belief that reality is grounded in such objects is a form of *objective* idealism. Plato's idealism is distinct, however, from the idealism of someone like George Berkeley, for while Plato accepted the, albeit imperfect and incomplete, reality of the sensuous world, as a reflection of the realm of Forms, Berkeley's *subjective* idealism posits that *only* minds and their ideas are real, and that sensuous reality is merely perception. It is Berkeley's idealism that characterises the extreme subjectivist approach to ontology, for subjective idealism requires the complete rejection of the material world, and therefore provides the clearest and most direct opposition to materialism—the ontological position we shall move on to consider next.

Concrete Particular

Directly opposed to the abstract universal is the concrete particular, which refers to spatiotemporal entities that embody instances of properties. Material objects are concrete particulars, and so too are particular events, having definite locations in space and time, and instantiating universals. Concrete particulars are grasped via the senses and have names that denote individuals. While universals apply plurally to objects in the world, particulars apply to just one thing at one time.

The concrete particular comprises the same part of the dialectical matrix that gives rise to the synthetic a posteriori in epistemology, and the connection between these is quite obvious. Both concern an external, experiential, and empirical reality, capable of interaction with and measurement by our tools of observation. The objects observed by the synthetic a posteriori are precisely concrete particulars, though the contents of empirical inquiry are nevertheless conveyed in terms of abstractions, such as relations, properties, and mathematical notation. As such, while the empirical process itself needs nothing more than concrete particulars, the enterprise of science would be hollow in the absence of abstracta.[80]

The belief that there are no such things as abstract objects is called 'nominalism', and its antithesis is Platonic realism. While nominalism provides a parsimonious solution to the problem of universals, the strain it puts on the language of science is seen by many as too great a trade-off. Despite this, both Charles Chihara and Hartry Field have argued that science *could* be done without abstracta like numbers, opposing Quine and Putnam's thesis that they are indispensable for it.[81] Quite reasonably, John Burgess responds that, "It is futile to urge a revolution in the practice of *physicists* motivated only by considerations appealing only to *philosophers* of a certain type," for there are no practical benefits in doing so, and it is unclear that it would even be possible for more abstruse areas of physics like quantum theory.

The ontological position that reality is composed solely of concrete particulars, and that the mental states we describe in common language do not refer to anything that is real, is called 'eliminative materialism' and comprises the extreme objectivist perspective on ontology. In combination with nominalism, the extreme materialist is held to the belief that all discussion is ultimately discussion of concrete particulars. Though we speak like objects have properties, and as though statements about properties, or numbers, or ideas, carry meaning, reality does not really have such propositional form, and the only true facts are facts about individual concrete entities.[82]

An antecedent to this perspective can be found in the atomism of the Presocratics Lucretius and Democritus, who were the first to espouse the view that matter is composed of individual particles called atoms. All change is a result of the evolving arrangements of these atoms, and qualities are merely conventions of thought, which is, itself, an arrangement of atoms. This variety of materialism has grown out of favour with the advent of 20th century physics, for the idea that everything consists of spatiotemporal particles leaves out much of our best descriptions of nature. Forces, for instance, are neither particular nor material, and particles like photons are massless and can occupy the same space. It is therefore more common nowadays to hear the term 'physicalism' in reference to the naturalistic perspective that reality is grounded in the elements of modern physics, but it is important to distinguish physicalism from the traditional metaphysical thesis that reality can be at least explicated in terms of concrete particulars.

Finally, *reductive* physicalists claim that certain abstractions, like beliefs or qualia, do exist in some sense but can nevertheless be reduced to, and are explicable in terms of, physical states in the brain. The task of this perspective is to show how mental states can be identified with physical states, in such a way that those states are not categorically and ontologically distinct from the physical, but not so indistinct that the resultant theory is less epistemologically informative from the stronger view of eliminativism. A less extreme form of physicalism still, *strong emergent materialism*, does allow for ontically novel mental states, and if eliminative materialism is the objectivist counterpart to Berkeley's subjective idealism, then strong emergent materialism has more in common with the *objective* idealism of Plato, with both positions having connections with substance dualism.

Abstract Particular

We now move on to consider the abstract particular, which is an entity without a determinate location in spacetime, and which is not a generalisation of a property. The most obvious candidates for abstract particulars are numbers or propositions, but on another view, the abstract particular presents a potential solution to the problem of universals—*trope theory*. A trope is a particular instance of a quality, such as a particular shade of the colour red, which is exemplified by a particular object. They are not spatially extended and are perpetually entangled with the state of observation, but they *are* localised to the particular objects being perceived.

Some trope theorists maintain that tropes represent merely the properties of the concrete particulars on which they depend,[83] but most hold that tropes constitute the fundamental ontic category from which other things are made. On the latter view, tropes are not properties of objects, but rather the objects themselves. This was the view of D.C. Williams, who coined the term 'trope' as "the alphabet of being" in the mid-1960s.[84] A precursor to this view can be found in the phenomenology of Edmund Husserl, who used the term 'moments' to refer to the individual instances of conscious experience, which form the basic elements of reality.[85]

In order for tropes to contribute both to the structure of the objective world as well as our experience of it, tropes must do a number of things. First, they must compose what we perceive as concrete particulars and abstract universals, but they must also inform us of what these things are like. If they succeed in this, they can offer a middle ground between Platonic realism—the view that things are like the universals in which they participate—and nominalism—the view that things aren't really like anything for they do not possess a propositional form. Tropes so conceived must therefore play the roles of both the object experienced as well as its properties, making each trope a trope of itself.

Furthermore, in order for tropes to satisfy our cognition of general qualities, they must inform us of how qualities can appear the same despite being different. Tropes can only be in one place and so the resemblance between two identical 'redness tropes' must be defined by its own 'resemblance trope', with the resemblance between resemblance tropes being defined via further tropes. Moreover, tropes must explain our sensations and intuitions of a concrete world, devoid of any concrete parts. Under trope theory, what we perceive in experience is simply a bundle of tropes, and so the Humean scepticism towards naive realism becomes dissolved. The particular instances of properties perceived *are* the object, and the mechanistic structure supposedly underlying them, which we do not perceive, does not exist until we do.

Considering a trope as both an object and a property leads to some difficulty, but no more than considering it as one or the other. Conceivably, a trope is neither an object nor a property yet something with the potential for both, reflecting the fact that the abstract particular expresses neither the subjectivist nor objectivist perspectives on the notion of being. On this view, tropes are mutable in the sense that the redness of a rose may cease and be replaced with some other trope, and indeterminate insofar as measurement no longer concerns an independent concrete world. Properties are dependent on what they are properties *of*, which is the thing being determined, whereas tropes are self-determining.

Perception plays an important role in trope theory, for it is only in the direct perception of properties that we are able to define and interpret phenomena. As Houston Smit states, Kant's 'intuitive marks' bear resemblance to abstract particulars and tropes, and how Kant interprets our access to tropes reveals one way that the abstract particular might be related to the corresponding class in epistemology—the synthetic a priori.[86] Kant states that an intuitive mark is a synthetic part of intuition because "it is a partial representation of a thing had in intuition", and is predicable of only one thing.[87] Our perception of the redness of a rose can be predicated as a property of that rose, where it is an intuition, or it can be predicated as an instance of the universal redness, in which case it is a concept.

A clearer connection can be seen if we take abstract particulars not as instances of qualities, but as mathematical entities like numbers. Again, modern foundations of arithmetic are grounded in synthetic axioms, in keeping with Kant's designation of mathematics as synthetic a priori. The belief that reality is fundamentally composed of mathematical or geometrical objects would be a form of Pythagoreanism—a view that has been recently revived in physicist Max Tegmark's *mathematical universe hypothesis*.[88] A further possibility is that abstract particulars comprise the substance of some fundamental plenum of pure information, in which case it would be a form of neutral monism, which holds that both the subjective realm of mind and the objective physical world are composed of a single class of elements, which are themselves neutral—neither mental nor physical.

Concrete Universal

The final possible combination of ontological categories is the concrete universal, which refers to an object which is at once spatiotemporal *and* something things have in common. The concrete universal comprises the same part of the dialectical matrix that gives rise to the analytic a posteriori in epistemology, which we understood previously to be innately contradictory. In a similar way, the concrete universal expresses that which is simultaneously contained in subjective experience and objective reality, resulting in the irrational synthesis of both the negative and positive aspects of complementarity—immutability and determinateness. Due to the mutual exclusivity of these properties, the monoletheic viewpoint struggles with the possibility of concrete universals being objects of our experience, for in terms of the dialectical matrix, the concrete universal is a contradiction of terms.

Despite this, there is a sense in which any monist can regard the concrete universal as an entity that surely does exist, and that is when it refers to the universal in which all particular substance is grounded, or as what substance itself is basically. Paradoxically, the limited mode of our experience prevents the concrete universal from ever being directly perceived, while also it must confront us persistently via the conduit of any phenomena. The concrete universal may be thought of as the whole of existence itself, or whatever the world fundamentally consists of, and is at once unperceivable yet constantly perceived.

Discussion of the concrete universal invariably arises in acknowledgement of Georg Hegel, for whom the notion was crucial to an understanding of individuality. Hegel has a more complex explanation for what makes a universal concrete, distinguishing it from the abstract in that, while the latter consists of the properties of individuals, the former consists of individuals themselves. The concrete universal is a universal of substance because it informs us of what its instances *are*, in their own individual ways, whereas the abstract universal informs us merely of what its instances are like, in terms of the property they exemplify.[89] Hegel is therefore using 'substance' in the Aristotelian sense of being a *genus*, or world, of individuals, which is differentiated and does not exist in abstraction from its instances.

Robert Stern, an expert on Hegelian philosophy, follows Hegel in giving the example of 'rose' as a concrete universal. The instances of the universal 'rose', being particular roses, exemplify the universal in their own individual ways, expressing different properties and abstract universals. In contrast, the instances of abstract universals are exemplified in the *same* way and are not therefore individuals. The concept 'red' is abstract because all red things are red in the same way, but also because its exemplars do not depend on their redness to be what they are; indeed, a white rose is just as much a rose as a red one. The difference is that not all things that are sometimes red are always red, but all things that are sometimes roses *are* always roses. Things are what they are by virtue of their substance, not their properties—by their concrete universal, not by the abstract universals they exemplify.[90]

The concrete universal is arrived at through a dialectical process that resolves universality and particularity within individuality, and indeed it is the relation between these categories that allow the dialectic its forward movement. Stern explains:

> Individuality is constituted by the particularized substance universal (as an individual, I am a man with a determinate set of properties that distinguish me from other men); the substance universal exists only in individuals, through its particularization (the universal 'man' exists in rebus, as instantiated in *different* men); and particularity is the differentiation of a substance universal, whereby it constitutes an individual (it is qua man that I have the properties that distinguish me from other men).[91]

Hegel's distinction between concrete and abstract universality depends on the idea that substantiality excludes abstractness. Yet, we do not consider a rose to be constituted of the universal 'rose', but rather of something which is itself abstracted from its properties, like matter or energy. Most things, even living things, are not roses, and so each species must have its own concrete universal. Yet, each species will itself be a particular kind of some higher genus, and the rose family is not concrete for its instances are not individuals. We must realise that Hegel's perspective on the concrete universal is inseparable from his perspective on the dialectic, and the dialectic is always heading towards an absolute perspective, which is the Absolute concrete universal. There are characterisations of the Absolute, such as 'rose', or 'person', or 'living being', but it is the self-reflexive *concept of* concrete universality which is the Absolute itself, for it is that which substantiates all other concepts.

Chapter Nine

On Value

Axiological Dichotomies

Our inquiry continues with an examination of the major concepts in axiology. This is a collective term for two distinct but interrelated topics: ethics, which is concerned with moral value or *virtue*; and aesthetics, which is concerned with artistic value or *beauty*. The relationship between axiology and ontology is somewhat similar to the relationship between ontology and epistemology, for the nature of existing things determines the sense in which values can be called real, and how they apply to the world. Nominalistic approaches to ontology preclude ethical and aesthetical value from possessing a mind-independent existence, while Platonic approaches suggest the converse.

In this inquiry, we shall focus our attention on the nature and status of *moral* values, which is the central concern of the area of study called 'meta-ethics'. However, the dialectical matrix can similarly describe meta-aesthetics as well as normative theories of value, which deal with moral *action*, and I will make some brief references to normative theories where it is useful to do so. Accordingly, the axiological categories that will be put to use in this section are derived from a pair of basic meta-ethical questions.

The first of these questions asks what it is exactly that we are referring to when we conceptualise the idea of value, goodness, or beauty. That is, providing we accept the existence of value in some sense, is that value dependent or independent of the mind perceiving it? If we contend that it is independent of the mind, then we are claiming that value refers to objective features of the world, that they are *real* things rather than mere beliefs or conventions, and that proclamations of value can be true or false insofar as they express moral facts.

If we posit that values are created by the mind, on the other hand, and are determined by attitude and opinion, then we are claiming that values are ultimately unreal. If we concede that value statements are meaningful insofar as they can be true or false, they are at least subjective.[92] Axiological subjectivism, and other cognitivist varieties of axiological anti-realism, thus hold that what makes a moral or artistic claim true is not how that claim relates to any mind-independent moral facts, but how the claim is to be viewed psychologically, culturally, or otherwise. That is, the subjectivist believes that a claim like 'killing is wrong' is not made true or false by any special facts concerning the act of killing, but rather by the disapproval of that act, which another person may not share.

One of the advantages of moral subjectivism is that, insofar as moral facts are grounded in facts about people, it allows for the possibility of using intuition and experiment to provide a normative theory of moral behaviour. On the other hand, insofar as people have different views on what is good and just, it allows for statements such as 'killing is right' to possibly be justified. A more general negation of moral objectivism is non-objectivism or anti-realism, which includes the more radical view that moral claims have no meaning at all, for they are related to the non-rational part of human experience and can neither be true nor false; this is called 'non-cognitivism'.

The first axiological dichotomy, which is an ontological dichotomy, thus distinguishes objective from non-objective notions of value, but if we accept that moral claims do express propositions, the dialectical matrix defines what kind of propositions they are, and so the distinction is between objectivism and subjectivism. It is the latter distinction that most clearly reflects the dimension of determinateness, on which the ontological dichotomy is predicated. This is because the mind-independence of value is inversely related to our ability to determine the truth value of axiological claims, from natural, subjective, or cultural conditions. In this sense, the two extremes of the distinction should be *non-naturalist realism* and *cognitivist anti-realism*, where realism and objectivism are synonymous.

The second axiological dichotomy refers to the applicability of value and is, therefore, more significantly normatively implicative. This dichotomy describes whether or not value judgements apply independently of time and circumstance—whether they are absolute or relative. A value is considered absolute if it applies universally throughout any context, such that the validity of a value judgement is immutable. For example, the idea that an action is always wrong, or that a painting is always beautiful, means that its value is absolute. If we posit certain things to be *inherently* good, then the value of those things must be derived from some immutable structure, whether that be the laws of nature or some system of ideal abstractions.

Conversely, a value judgement is considered relative if it has truth relative only to certain social or subjective conditions. For example, a painting may be beautiful to one person but ugly to another, an action just to one person, unjust to another. Such values are mutable because their truth and applicability change with changing circumstances, and so the moral relativist believes that when it comes to proclamations concerning value, one is not *necessarily* right or wrong.

These are thus the four basic categories of the notion of value. The immutable aspect is represented by the absolute because it refers to a fixed and unchanging standard of judgement, while the mutable is represented by the relative because it refers to judgements that depend on circumstance. The indeterminate aspect is represented by non-natural objective values as they are not extractable from any amount of study of the natural world, and the determinate aspect is represented by the subjective, for such values are reducible to non-moral conditions such as mental states or cultural standards, which can be studied scientifically.

We can now represent these categories on the dialectical matrix to produce four classes reflecting the possible cognitivist descriptions of value. The matrix *could* be expanded to include non-cognitivist theories by exchanging the objective–subjective dichotomy for a real–irreal dichotomy, but since non-cognitivism, as well as nihilism, are forms of scepticism towards the truth of moral claims, they are not required to be expressed by the dialectical matrix, which is concerned with positive theories. The four classes of value about which we can meaningfully reason are thus the *objective absolute*, the *subjective relative*, the *objective relative*, and the *subjective absolute*. We continue with a description of these classes.

	Objective	Subjective
Relative	Objective Relative Pluralistic	Subjective Relative Relative
Absolute	Objective Absolute Objective	Subjective Absolute

Objective Absolute

The objective absolute refers to values that exist independently of the mind, as real features of the world, and which apply universally across all time and context. For example, the idea that killing is always and necessarily wrong, due to the inherent nature of the act, is based on an objective and absolute value. One might argue that killing is absolutely wrong because 'wrongness' is contained in 'killing' analytically, and that there is no conception of killing that is not wrong.

The belief that values are objective and absolute encapsulates the subjectivist perspective on axiology and comprises the same part of the dialectical matrix that gives rise to the abstract universal in ontology. The classical example of objective absolutism, which expresses this connection, can be found in Plato and his theory of Forms. For Plato, values exist in the form of abstract universals, such as the Form of Justice, of Virtue, and, most crucially for Plato, the Form of the Good. A thing is made good insofar as it participates in the Form of the Good, which is the highest expression of the Forms as it is 'the light' which illuminates all those lower Forms and allows us to gain knowledge of their nature.[93] If a Form is a perfect version of an idea, then the Form of the Good is the perfection of perfection itself.

In Plato's ethics, values cannot be reduced to any natural features of the world, and objectivism is often taken as an anti-naturalist theory. This view was made clear in 1903 by G.E. Moore's 'open-question argument', which argues that *if* values were grounded in some natural property—for instance, if goodness was merely an expression of pleasure—then the fact that a pleasurable act is a good act would be an analytic necessity. Given that it is perfectly reasonable to question whether something is good, or virtuous, or just, such moral concepts cannot be identified with natural properties. If values exist, they must be something over-and-above the natural domain.[94] Moreover, in regard to absolutism specifically, a naturalist would have to claim that moral imperatives supervene on facts about human nature, such that they apply equally to all people.

Since there are inevitably people who are either unaware of these imperatives or just don't care to observe them, it's hard to say that they apply equally to all people.

It is also in its non-naturalistic form that objective absolutism expresses the properties of immutability and indeterminateness, for if values cannot be derived from natural conditions, the question arises as to how we can gain knowledge of them, and how they apply to the world. If they are absolute, then they must have universal authority, but this does not entail that we have any means of developing an effective normative theory of them. It may come as no surprise therefore that many absolutist systems of ethics are transcendentalist or religious, whereby the central texts of those religions—the Bible, the Torah, the Quran—are taken as a divine source of moral truth. Take for example the Golden Rule—treat others as you yourself wish to be treated—which is found in some form within the vast majority of the world's religions.

Normative absolutism need not be ordained by some transcendent authority, however, for there may be certain moral principles that can be asserted on the grounds of a priori reasoning, thus applying to all rational agents equally. One such approach is found in Kant's 'categorical imperative', which asserts an action to be moral insofar as it would be a good thing for all people always to undertake that action—that one should "act only in accordance with that maxim whereby one may will that it should become a universal law".[95] The imperative is considered categorical because defying it would invoke a contradiction in one's own will. That is to say, only those maxims which do not contradict our conception of goodness, when universalised, are categorically imperative, regardless of the consequences of adhering to them. There is some debate, however, as to whether Kant's ethics should be considered a form of moral objectivism, for at least as far as the imperative is concerned, moral principles are grounded in rational agency.[96]

Subjective Relative

Opposing the objective absolute is the subjective relative. This class refers to values that exist relative to circumstance, do not exist independently of the mind, and are reducible to facts concerning human nature. For example, the idea that killing can be morally justified insofar as the perpetrator has deemed it just, is expressive of subjective relative value. Here, killing is not inherently wrong, and so it is a synthetic process that binds the concept 'wrong' to the concept 'killing'.

The subjective relative comprises the same part of the dialectical matrix that gives rise to the concrete particular in ontology, and these two are related in the sense that a world consisting only of concrete particulars will not contain non-natural or objective value. Any values that do exist, and which can be predicated of that world, must exist relative to human nature. The belief that values are subjective and relative encapsulates the objectivist perspective in axiology—moral relativism—though non-cognitivist varieties of anti-realism also fit seamlessly into an empiricist or materialist paradigm.

The earliest theory of subjective relativism was given by the Presocratic philosopher Protagoras, who purported that each individual possesses their own unique and equally valid notions of goodness, and that people should do whatever it is that is in their best interest. Another important figure in the history of relativism is David Hume, though Hume's position is more accurately regarded as a precursor to modern emotivism—a non-cognitivist theory. For Hume, human reason is a "slave of the passions", and so ethical behaviour is merely a consequence of the way that people feel.[97] As such, when we talk of concepts like virtue and beauty, we are referring to people's attitudes and emotions; that is, rather than being an intellectual matter as to whether a particular action is wrong, moral judgements are purely conative.

In contrast to the moral rationalism implied by any normative theory of objective absolutism, morality based on emotion, motivation, sentiment, or biology is uncovered via empirical inquiry into human nature and is therefore relative to these factors. Such moral sentimentalism implies the mutability of values, insofar as circumstance determines the justification of value judgements, and determinateness, insofar as the factors determining sentiment are objective and measurable.

Objective Relative

The third class of value is the objective relative, which refers to values that exist as real features of the world, but where context is also a determining factor on their applicability and the validity of moral claims. Traditionally, objectivism and relativism have been presented as opposing theories, and it is often taken for granted that when one speaks of relativism, they are also referring to anti-realism. Nevertheless, each of these refers to a separate dimension of axiological thought, and the positioning of objective relativism on the dialectical matrix suggests that it is a perfectly feasible position. In addition, theories of objective relative value have grown increasingly popular over the past 100 years, for they offer some recourse to the intolerance of absolutist moral codes, without being forced to admit of pure subjectivism.

There are a number of ways that we can approach objective relativism, and the simplest is to say that while some judgements, actions, or moral claims are governed by absolute and inviolable principles, the status of others is relative to circumstance. A more satisfying approach is to say that in any given circumstance, there is a multitude of relevant objective facts regarding value, and there is a wide range of valid opinion that falls under the specification of 'the moral'. As such, moral truths act like guiding principles rather than imperatives, and different interpretations of value may be more or less applicable at a given time than others.

Coming from the perspective of moral relativism, David Wong has argued that while moral truth is determined by purely objective factors concerning the human condition, there is nevertheless no singularly true ethical code, such that what is good and just in one society may well be deficient in another.[98] In this way, given that there are differences in human needs, wants, and societal challenges faced, there are accordingly different and equally justified moralities. There are, however, limitations to what can and can't be a true moral claim, guided by a *common* good, so that values are not subjective.

Latvian-British philosopher and political theorist Isaiah Berlin reiterates this point with a focus on the values themselves:

> There is a plurality of values which men can and do seek, and these values differ. There is not an infinity of them: the number of human values that I can pursue while maintaining my human character is finite...If a man pursues one of these values, I, who do not, am able to understand why he pursues it or what it would be like, in his circumstances, for me to be induced to pursue it.[99]

Wong affirms that our disagreements with others are seldom tied to a belief that those people are irrational in their judgements, and that such 'moral ambivalence' hints towards a pluralism of values.[100] Berlin's 'value pluralism' asserts that there may be many values that apply to a judgement in different ways. In one circumstance, a single value may apply more so than others, while in another, there may be several values that apply equally, allowing for several equally valid judgements. These values are objective since they emerge consistently throughout different cultures, suggesting to Berlin that they are "part of the essence of humanity rather than arbitrary creations of men's subjective fancies".[101]

Normatively, value pluralism can be used as the starting

point for Aristotelian 'virtue ethics', where different virtues—kindness, patience, forgiveness—may come into conflict. Alasdair MacIntyre argues that the diversity of importance placed on virtues in different cultures means that there can be no universal moral code.[102] He seeks to revive the Aristotelian notion of *telos*—of our intended end goal—the decline of which following the Enlightenment period signals to him the demise of modern moral philosophy. The difference is that universal moral codes, be they deontological or consequentialist, prescribe virtues *based* on their code, while virtue ethics views the virtues themselves as the starting point for moral action.

Finally, the objective relative comprises the same part of the dialectical matrix that gives rise to the abstract particular in ontology, and a connection can be found between them if we accept that there is not a single ideal of a recognisable value. That is, rather than there being a *thing* called 'virtue', there are simply *things* that are virtue-*like*, whereby virtue is defined as the resemblance between these things. Whereas the Platonist can hold that objects and events partake in absolute axiological ideals, and where the nominalist can hold that they accrue value conferred by minds, the trope theorist might maintain that particular properties are the things that are valuable, rather than the objects as a whole. The value of tropes would then apply relativistically towards the value of the compresence of those tropes within the object or bundle.[103]

Subjective Absolute

The final possible conjunction of axiological categories is the subjective absolute, and this class refers to values that exist dependently on the mind, as attitudes or opinions, but which also apply universally and absolutely across cultures, times, and circumstances. The subjective absolute comprises the same part of the dialectical matrix that gives rise to the concrete universal in ontology, which we understood previously to be phenomenologically self-contradictory. Similarly, the subjective absolute represents a perspective which is epistemically subjectivistic, in the sense that it considers values to be certain and immutable, but also objectivistic, in the sense that it considers values to be limited to, and determined by, natural conditions.

Due to the complementarity of immutability and determinateness, we would not expect it to be possible to develop a method of prescribing moral action in regard to this view. If we allow that value depends on a mind, then moral judgements are justified whenever the subject considers them to be so. However, insofar as individuals are bound to disagree on the moral worth of actions, we are led to conclude that two subjects may be equally justified in conflicting beliefs, thus entailing relativism. This is therefore to say, if we want values to be absolute, they cannot then be grounded in human opinions whenever those opinions feature variance. Only if it were the case that everyone always agreed on moral judgements could normative subjective absolutism obtain, in which case we would need only to investigate what we all agree on to determine axiological truth.

Following on from this observation, some thinkers, such as Roderick Firth, have considered a form of moral subjectivism whereby the subject involved is not the producer of a judgement, but a hypothetical observer that is fully informed of all relevant facts, is immune to any kind of bias, and has an infallible capacity for reason. According to this idea, it matters not that people disagree on what characterises virtue or beauty, because people are fallible. There may indeed be an absolutely correct judgement or action for any given circumstance, but this can only be ascertained by such an 'ideal observer'.[104]

That valid judgements are those that would be approved of by an ideal observer does not, of course, provide us with any means of determining what those judgements are, for unfortunately we are yet to meet such an observer in the world. 'Divine command' theorists might contend that God is such a subject, but without empirical access to perfectly informed judgements, a normative theory of subjective absolute value remains untenable. Moreover, the meta-ethical theory seems to fall victim to circularity problems. Since an ideal observer can only be defined as one who makes proper value judgements, and 'proper value judgements' can only be defined as those judgements an ideal observer would make, there is no way to learn what moral claims are true without simply knowing what moral claims are true. Judging in accordance with the subjective absolute therefore requires the exercise of a function that all too often we do not use, though surely sometimes we do.

Chapter Ten

On Right

Politico-theoretical Dichotomies

As we embark upon this final inquiry, let us be clear that this is a study of the basic perspectives from which we can approach political theory, and not about which of these perspectives might be better than others. Accordingly, we shall focus on the basic value systems on which these approaches are grounded, giving little attention to the practicability of respective political doctrines. Among all the topics of philosophy, political theory in particular is one we are unavoidably faced with, and, understandably, opinion is rife. In any theoretical inquiry, which also implies a praxis, we must, if we wish to be impartial, acknowledge the difference between why political ideas are conceived of in theory, and how they end up being realised in practice. There is what we believe is right, and there is also what works; these two things need not be in alignment, and it is much easier to conceive of an idea as an individual, than to implement it as a collective. In the present study, we are principally concerned with the ideology underlying the theory, rather than its practice.

Political theory is, in principle, built upon axiology, for the way that values are realised at the social and economic levels determines the legitimacy of governance and law, and the urgency of duty and right. We shall therefore discuss the relationships between meta-ethical and political categories with a little more care than in previous sections, as this will help us to understand how political theories are expressed within the underlying matrix of philosophical thought. It will be useful to note at this point that the complementary properties of *immutability* and *determinateness* can be translated into the socioeconomic rights of *equality* and *liberty* respectively—a correlation that will be elucidated upon as we move forward.

The first dichotomy that we shall consider is concerned with personal freedom, and the legitimacy of any government in restricting that freedom. It is therefore a *social* dichotomy that distinguishes the importance placed on the responsibilities of the state, from the individual's right to liberty. Ideologies which place more importance on the former tend towards *authoritarianism*, and the latter towards *libertarianism*. Of course, the authoritarian may consider some authorities more worthy than others, and the legitimacy of government is ultimately subject to the opinions of the governed. Here, we will interpret legitimacy in a moral sense, whereby an authority is legitimate when it acts on behalf of what are seen to be inviolable principles of social value, whether utilitarian, deontological, or otherwise. A legitimate authority is grounded in principles that are not solely political in nature, and so it does not legislate purely from the desires of its members or subjects. It is an organised manifestation of axiological principles, around which laws are enacted to ensure their adherence and acknowledgement in society.

As such, authoritarianism can be seen as reliant on the *objectivity* of moral value, and a similar connection can be seen in a variety of disciplines outside of politics as well. For instance, the objectivity of scientific truth gives that study a form of authority over those ideas with which it conflicts—an attitude that is common in the dialectic between the atheist and the religious fundamentalist. Under moral objectivism, law may reflect ethics insofar as punishment towards the insurgents of law is justified if and only if certain actions are objectively wrong, or if they produce results that are objectively good.

In her 1991 book, *Contingencies of Value*, literary critic Barbara Herrnstein Smith attacks moral objectivism on the *basis* of its authoritarianism. She cites objectivists like Immanuel Kant and

Alasdair MacIntyre as being held to the belief that the objectivity of value means that certain societies are better or greater than others. This, she feels, inevitably translates into social hierarchy and inequalities of power, as those individuals who are out of line with the objective values are degraded to a lower status in society. Robin West states in her critique of this work, "Smith argues that all objectivist theories of value—by which she means those theories which ascribe objective value to the thing being valued—in both aesthetic and moral philosophy...are irretrievably authoritarian."[105]

The idea that moral objectivists are inherently less capable or willing to appreciate the diversity of people is not unreasonable, but that this means they are "irretrievably authoritarian" is undoubtedly an unfair generalisation. Political intentions in the real world are not solely ideological, but also practical; better to say, as Robin West does, that the evaluative reasoning of the objectivist "is always *in danger* of becoming authoritarian", but not *necessarily* so.[106] This danger is well studied in the psychological literature, where "political conservativism has been found to be associated with greater tolerance of inequality, lesser tolerance of change, greater conscientiousness, and greater sensitivity to disgust".[107]

On the other hand, liberal ideologies have been found to express greater empathy and openness to change, as well as having a more subjectivistic perspective on the notion of moral value.[108] This leads us on to the antithesis of authoritarianism—libertarianism—which upholds individual liberty as its core principle, rejecting the legitimacy of an authority to encroach upon this liberty. Libertarianism arose in public thought during the Enlightenment period in Europe through the work of thinkers like John Locke, who argued that governments exist solely for the benefit of individuals, and that therefore the liberty of the individual should be held at the highest regard. Notably, liberty is here defined in a negative sense, as freedom from interference from others, as opposed to positive liberty—the freedom to do whatever one pleases—which, as Isaiah Berlin has expressed, implies a need for government intervention.[109]

Libertarianism at its extreme is synonymous with anarchism, whereby centralised authority is absent entirely. All forms of libertarianism, however, reject the idea that someone's individual needs impose any kind of moral duty on others, and so as long as the individual rights of people remain unviolated, there is no justification for inhibiting any actions they might undertake. It is here that we find the connection between libertarianism and moral subjectivism, for the libertarian does not want anyone's subjective point of view to be *objectivised* and imposed on others. Libertarianism has also been attacked on the basis that it implies some degree of nihilism and the renouncement of common values[110]—an idea that is also familiar to fiction, an example being found in Dostoevsky's *Demons*.

The second dichotomy of political theory concerns the economy—how it should be run, who it should be for, the freedoms and limitations placed on it, and its relation to public enterprise. It is the economic dichotomy that distinguishes the left-right spectrum that is common to public discourse. It is often taken for granted that left-wing politics implicate higher levels of governance and authority, while right-wing politics does the converse. This is, however, merely a result of the interaction between the economic and social dimensions of civil life, and in a purely economic sense, the left versus right distinction is largely oriented around how much the individual will is valued in relation to collective society.

The *collectivist* perspective, on the left, holds that the economy should be a cooperative enterprise, regardless of whether that be state owned or communal. Collectivism is the belief that what is best for whole groups and communities is more important than what is best for individuals. It is grounded in the notion that there is a common or universal set of values that apply equally to all people. The collectivist may be more willing to dispose with individual rights, if those rights pose a threat to the common good, and so collectivism is connected to axiological absolutism. This is to say, in a society based on collectivism, everyone adheres to a common set of values, and this common set of values derives from a moral paradigm which is innately universalist.

Modern collectivist thought was mobilised in 1762 by Jean-Jacques Rousseau's *The Social Contract*, which would influence political reforms in France and elsewhere in Europe. Rousseau believed that the most decisive facet of our evolutionary history is the ability for cooperation, which allows us to achieve things no individual ever could. Accordingly, individual freedom is realised, for Rousseau, only via submission to the 'general will' of the collective society. Adam Smith—a contemporary of Rousseau—took a different view; he believed that society flourishes when each individual seeks only their own advantage, and when economic life self-manages in accordance with the natural tendencies and talents of each of them.

Smith's *individualism* was inspired by the 17th century philosopher Thomas Hobbes, who professed a radical psychological egoism. For Hobbes, humans are machines relentlessly driven by their own self-interest, and are therefore unwilling and unable to cooperate. However, while Hobbes took this to require the subordination of individual rights to the power of a just authority, and thus to collectivism, Smith saw the self-interest of individuals in the positive. It is on this principle that the foundations of capitalism were laid, where labour is the unit of value, where progress arrives through competition, and where the markets are free from government intervention.[111]

Individualism is tied to moral relativism for if individuals are encouraged to pursue what they find most valuable, there cannot be an absolute standard that applies equally to us all.

Furthermore, the moral worth of individuals may be relative where their value depends on attributes attached to them, such as those which increase productivity. Individualism holds that the freedom and desires of individuals surpass any alleged common good, wherein individual differences are venerated and the right to autonomy is sanctified.

And so, we have now identified the four basic categories of political theory, being socioeconomic expressions of the complementarity of immutability and determinateness. Collectivism reflects the property of immutability in its emphasis on the uniform aspects of society, such as the fact that individuals are all equal in their humanity, and in political theory this is expressed as *equality*. Individualism reflects the property of mutability in reference to the variable aspects of society, which is expressed by individual differences and *inequality*. Authoritarianism reflects the property of indeterminateness in the restriction it places on the self-determination of individual autonomy, and this manifests as a presence of *authority*. Finally, libertarianism reflects the property of determinateness in the

	Authoritarian	Libertarian
Individualist	Authoritarian Individualist Conservative	Libertarian Individualist Capitalist
Collectivist	Authoritarian Collectivist Socialist	Libertarian Collectivist (Communist)

same way as subjective values, or in the freedom for individuals to pursue their goals without restraint, expressed as *liberty*.

As with all previous cases, these four categories form two dimensions which interact with one another to form four distinct classes of socioeconomic perspective. We continue again with a brief description of each.

Libertarian Individualism

Libertarian individualism, right-wing libertarianism and *libertarian capitalism,* including both *minarchism* and the more extreme *anarcho-capitalism,* place the moral worth of social liberty over that of economic equality, accepting the consequence of inequality sometimes even with favour. This manifests as little or no redistribution of wealth via taxation, privatised public services, private ownership of the means of production, and a competition-based free-market economy.

Minarchism permits some of the infrastructures of government as a means of protection, such as military defence and law enforcement, but anarcho-capitalists argue that the existence of any state violates the individual's right to liberty.[112] They believe that systems of compulsory taxation constitute the definition of theft, and they want all public enterprises to be privatised in a market absent of politics. Murray Rothbard forwards this view on the grounds of a principle of non-aggression, whereby the sovereign ownership of one's person means that no individual, state, or collective has the right of "aggressive coercion" over another.[113]

Rothbard's approach is based on an absolutist and deontological view on human rights, though libertarian individualism comprises the same part of the dialectical matrix that gives rise to subjective *relativism* in axiology. Other libertarian capitalists, such as John Stuart Mill, Ludwig von Mises, and Friedrich Hayek, offer a consequentialist approach that is indeed based on a subjective and relative definition of values, and ethical egoism is also commonly presented as a moral ground for individualist anarchism.

Authoritarian Collectivism

Dialectically opposed to libertarian individualism is *authoritarian collectivism*, encompassing *left-wing authoritarianism* and *authoritarian socialism*, or socialism-from-above. Where the former values above all else the individual right to liberty, authoritarian collectivism allies itself with economic equality. This discriminatory loyalty to equality necessitates the authority of a state, which controls the means of production, finances public services, and redistributes the national income.

In considering what would be required for the realisation of equality, we must first define what that equality should be of. Early meritocratic societies, such as that of ancient China, were fully hierarchical yet had equality in a sense. Meritocracy, where power is held by those of greater *merit*, can have an equality of opportunity, but not equality of outcomes, and it is the latter which we usually consider as constituting a positively equal society. How such equality can be achieved between an authority and the proletariat is not a simple matter, though socialism generally is focussed on equality of access to the means of consumption.

The clearest example of authoritarian socialism comes from Vladimir Lenin, who saw the most effective route to a classless society to be via the implementation of a vanguard state that would rule on behalf of the working class.[114] Karl Marx, contrarily, envisioned communism as the natural evolution of a highly developed capitalism, in which the public themselves would revolt and overthrow the bourgeoisie.[115] It has nevertheless been argued that collectivist systems entail authoritarianism by necessity, since to implement equality power must be instilled in some form. Friedrich Hayek, for instance, argues that collectivism implies the valuing of the state above that of the individual, and that this makes possible the emergence of totalitarian regimes.[116] The danger of totalitarianism is also implied by the fact that authoritarian collectivism comprises the same part of the dialectical matrix that gives rise to objective absolutism in axiology. This is because if the value of a state has an absolute standard, then the defence of this standard *by* the state may preclude the democratic process.

We commonly recognise the monopolisation of political power to be associated with tyrannical behaviour and the promotion of politically motivated propaganda. However, there is no theoretical reason why such a state could not be associated with a legitimised process and be disengaged with any self-interested ideology. Historically, this is evident in the divine right of kings, which gives legitimacy to the absolute rule of a monarch, and many religious institutions, such as the papacy and the succession of lamas, are undoubtedly totalitarian while also highly respected and culturally important.

Authoritarian Individualism

Authoritarian individualism, encompassing *right-wing authoritarianism* and *authoritarian capitalism,* is comparatively the most applicable political class to the modern developed world. It refers to political systems that value obedience to, and support from, an authority as a means of maintaining the social structure, and which is supported via taxation within an open and competitive market. It is a moderate class in relation to the previous two, for it is devoted neither to the extreme of social liberty nor to that of economic equality. While the libertarian and the socialist may be led to conflict over the moral superiority of these values, the authoritarian capitalist abstains, instead seeking to achieve some balance between them.

In the dialectical matrix, this is expressed by the fact that authoritarian individualism is absent of both immutability and determinateness—a position it shares with objective relativism and value pluralism in axiology. Isaiah Berlin's pluralism holds that there are objective values, but that these values conflict with each other in such a way that there may not be a completely correct course of action one might take. Crucially, liberty may conflict with equality, and we may be required to make choices that violate one or the other. Berlin took the necessity and legitimacy of free choice to provide the ground for his conservative variety of liberalism. A summary of the view developed by Berlin in *Two Concepts of Liberty*, taken from the Stanford Encyclopaedia of Philosophy, reads:

> Berlin was more sensitive than many classical liberal or libertarian thinkers to the fact that genuine liberty may conflict with genuine equality, or justice, or public order, or security, or efficiency, or happiness, and therefore must be balanced against, and sometimes sacrificed in favour of, other values. Berlin's liberalism includes both a conservative or pragmatic appreciation of the importance of maintaining a balance between different values, and a social-democratic appreciation of the need to restrict liberty in some cases so as to promote equality and justice, and to protect the weak against victimisation by the strong.[117]

Berlin's conservatism is of a distinctly practical nature, for in any society we find a wide array of political perspectives. Authoritarian individualism is not merely a pragmatic means of maintaining balance, however, but a very real outcome of the democratic process, in which there are individuals who would wish to defeat this balance. Those steadfast on liberty pull the system away from equality, and those on equality away from liberty. Conservatism in general is quite clearly associated with authoritarian individualism, particularly in its more reactionary aspects, such as the desire for security, stability, tradition, and a return to a past state of society. These are all things that justify the need and desire for a powerful centralised authority.

There can also be much more radical varieties of authoritarian capitalism. In the extreme, the economy at large begins to look like one huge corporation that is owned, managed, and insured by the state. A step before this, as a collection of monopolies and oligopolies, which are actively protected by government policy, and where taxpayers are called upon to bail out institutions during times of crisis. 'State capitalism' is the name given to an economy that has nationalised the means of production, and China following the Mao era, from 1978 until present, is our best example.

There are varying degrees of government intervention in the economies of the developed world, but the majority of the liberal politocracies of the West fall squarely within the class of authoritarian state-oriented capitalism. Libertarian socialist Noam Chomsky has also commented on how large-scale corporations in the United States are effectively insured and protected by the government since they receive public money during bailouts. The risk these businesses assume is therefore greatly reduced at a private level, and Chomsky takes this as a form of nationalisation of industry.[118]

Libertarian Collectivism

The final class for us to consider is that of *libertarian collectivism*, which here shall be taken to encompass *libertarian socialism* or socialism-from-below, *anarcho-socialism*, and *left-wing libertarianism*. The libertarian collectivist society is oriented both by the freedom for self-governance *and* by the moral worth of the collective. It expresses the basic tenets of Marxism, and classical Marxists regard it as the only genuine form of communism. It thus represents the end-stage of the transition from capitalism to communism, wherein widespread public ownership of the means of production, as well as equal access to the means of consumption, has been achieved. There is also an absence of the state and social classes, and so classical Marxists reject the vanguardism employed by the Marxist-Leninist movements of the 20th century. Instead, they advocate for 'revolutionary spontaneity' as the proper path to communism, in which a proletarian uprising self-organises from the desires of the working class, without any direction from a *professional* socialist agenda.

Nevertheless, the fact that libertarian collectivism seeks the simultaneous exaltation of both liberty and equality is a matter of great contention. Some hold it to be the true and necessary goal of civilised man, and others as a hopeless contradiction. In terms of the dialectical matrix, libertarian collectivism shares with authoritarian individualism a neutrality between equality and liberty. While the authoritarian individualist sees the two as incompatible and antagonistic, however, the libertarian collectivist views them as wholly commensurate and complementary.

The distinction between these two neutral ideologies expresses a dialectic between optimism and pessimism in regard to human nature, though optimism towards the possibility of libertarian collectivism has been proved to be a naive view. The libertarian collectivist can defend themselves, however, by arguing that human society is similar to all other self-organising processes—the evolution of species and the evolution of language—and that our ability to generate harmony in the midst of conflict is a slow but natural process that we are just not yet advanced enough to see.

To what extent liberty and equality truly are antithetic has been a recurrent topic in the literature. Their incompatibility is implicit in the socialist's willingness to sacrifice some liberty, as it is in the capitalist to do the same with equality. Surely, we would prefer capitalism *with* equality, or socialism *with* liberty, if possible. However, if all people are equal then there must be some restrictions on freedom which have facilitated that equality, and if all people are liberated, they shall be free to accumulate more power and wealth than their neighbours. Any organised attempt to redistribute excess wealth, either by coercion or by force, would constitute a reduction in liberty, and so the two rights do seem to be striving toward entirely conflicting ideals.

However, what this reasoning leaves out is the possibility that a restriction on freedom could be one of self-restraint and will, rather than one of force—that the redistribution of wealth could be self-motivated from the desire of each individual to be equal. In a society where each individual *wills* both liberty and equality, and where they act in accordance with their will, there is no *logical* contradiction between either. Nor would it be the end of individualism, for the motivation to sustain the system would originate from each individual. This, in turn, requires the recognition of a common good, but not one that is authorised, learned, and which should be *obeyed*. Rather, such a common good must be purely self-interested, as expressed in the intuitive distinction between ethical behaviour based on obedience to objective rules, and ethical behaviour for the sake of ethical behaviour.

Whether we shall ever realise such moral will is a matter of great contention, and this facet of libertarian collectivism reflects its axiological counterpart within the dialectical matrix—subjective absolutism. As we understood previously, the prescription of subjective value implies relativity unless everyone always agrees. Similarly, in a truly democratic society, where the people themselves rule, equality must be maintained through the self-organised cooperation and charity of individuals. Libertarian collectivism *relies* on each individual being an ideal citizen—an ideal observer—who understands that what is good for the collective is good for the individual, and never strays from this principle.

Such a hypothetical society (and it is *very* hypothetical) could even achieve liberty in a positive sense, for while each individual may be free to indulge without limit, they may choose by their will not to do so. There is a very real sense therefore in which liberty and equality, far from being a contradiction of terms, are nothing other than perfect complements. To boot, how can we be *absolutely* equal when we have less power than others, and how can we be *positively* free whenever we have lesser means. The fact that both are complementary, not contradictory, only makes them more difficult to realise, of course, but it is hard to deny that the simultaneous expansion of both liberty and equality is a noble and worthy goal for humankind.

This is the view of the optimist, and we must realise too that the pessimist has an equal claim to truth. On their view, libertarian collectivism is a utopian fantasy that disregards the harsh reality that human beings will never transcend their selfish ways, and that sacrifice is incessantly needed. Most of us in the West live in largely neutral and centrist societies, though many would loathe to accept it. We have not liberty nor equality, but this is not the result of some grand conspiracy, it is simply the only way a democracy can be, representative or otherwise, now, at this current stage in our evolutionary history.

There are people everywhere of all views and bents, all reacting to each other's excesses. The only medium between the two conflicting spirits of society is naturally one of centrism. The defenders of liberty and equality respectively will never give up on their visions willingly, and so civilised societies are inevitably and irretrievably *forced* to give up on both, with the consequent of authoritarian individualism. It is an unavoidable reality that wheresoever a democracy cannot discriminate between the good of liberty, and the good of equality, so shall that democracy have neither. What it will have, though, is equilibrium, for the harder we push in one direction, the greater rebellion will be felt in the other.

Chapter Eleven

Introduction to Syntheorology

A Geometry of Conception

Now that we have ascertained with some certainty that there is a uniform structure underlying the concepts and theories of the four major branches of philosophy, we can introduce a methodology to the dialectical matrix. The result is a subject I call 'syntheorology', being a study of the relations between distinct philosophical ideas.

Syntheorology will prove to be a useful tool in our quest for knowledge in a number of different ways. First and foremost, it will provide a framework for the study of the basic properties that constitute the content of a given philosophical theory and the underlying matrix from which these properties and theories emerge. In this sense, syntheorology is a kind of geometry of the reasoning capacity, being wholly unconcerned with the effectiveness of our claims and caring only of the underlying structure from which they arise. In other words, where the object of philosophical inquiry is true or good ideas, the object of syntheorology is the source of all ideas—the dialectical matrix—and the process of translation from pure imagination to the categories of thought.

Syntheorology is not, therefore, a theoretical or hypothetical inquiry, and does not aim to justify nor vilify any theory. It is a descriptive discipline that develops the universal which binds all particular views, and it is indifferent to the monoletheic assumption that there is a single theory that answers all philosophical questions. It is concerned with whatever it is at the bottom of the cup when it comes to our ability to *attempt* monoletheic philosophy—our basic capacity to envision *something* as being true, and the dialectical matrix *through which* this capacity is translated into the language of concepts.

The dialectical matrix is integral to syntheorology because it is the framework that determines what kind of theories are possible. In two-dimensional geometry, a matrix can be thought of as an invisible structure, like a grid, that underlies the surface upon which images are grounded, and the structure of the matrix determines the patterns that can be imposed upon it. For example, an electronic screen is composed of a matrix of pixels, and any image displayed on a screen is dependent on the structure of those pixels. In this way, there are no true curves on a screen, and any appearance of curvature depends on the size of the pixels in relation to an observer. Loosely speaking, the dialectical matrix can be thought of as that which binds the fluid and continuous nature of our capacity to believe, into the discrete and definite forms of our ideas.

Abstractly, matrices can also provide the foundation for continuous patterns, such as an overlapping grid of circles. One example of such a grid has been referred to as the 'Flower of Life', and from this structure every possible regular convex polyhedron can be produced—the *Platonic solids*. We can make an analogy here of the distinction between philosophy and syntheorology, whereby competing philosophical theories are equated with the polyhedra, and where the intellect itself is equated to the Flower of Life. In philosophy, we identify these shapes through reason, choose which one we like the most, and then argue that *our* polygon is exclusively right and just, the others being mere delusions. In syntheorology, conversely, we recognise all those different shapes identified by philosophers, and develop scientifically the matrix which defines them. The polygons themselves—our theories, concepts, and beliefs—are a means to an end, serving only as an excavation of our common sense, and of the dialectical matrix which guides it. Syntheorology works from the outside in; rather than following one path down the rivers of consciousness to arrive at some conclusion—to some conception we postulate as true—we follow all rivers upwards to the source they share together—to that capacity within us which generates the notion of truth.

An Archaeology of Intellect

Syntheorology can provide a powerful methodological tool in that an understanding of the geometrical relations of ideas can allow us to derive new ones without the use of philosophical reasoning. Typically, a philosophical theory can be considered effective insofar as it is reasoned with skill. A good theory resolves existing problems while avoiding new ones, is comprehensive yet parsimonious, is technically proficient and presented in accepted jargon, and so on and so forth. These things all implicate at least two important factors. First, they rely on our ability to reason well; and second, they require theories to be presented in such a way that they are *better* approximations of truth than others, and so there is implicit conflict in any rhetorical efforts to defend a given idea.

The entire enterprise of theoretical philosophy can be seen to be grounded in a very particular way of thinking, which is fuelled by contradiction in the sense that if one idea is good, then its antithesis must be bad. When combined with monoletheism, these good and bad ideas are seen as attempts to imitate truth and falsity, and there is a clear division between them. This methodology is just if it shares its logic with the structure of reality, but to assume that one's reasoning is wise because it describes a world in which one's reasoning is wise, is a vicious circle.

Syntheorology offers us a new method and motive for developing theories—one that is not based on the partial, affective, and subjective faculties of human reason, but on the basic properties and relations of the underlying matrix on which they supervene. We might then question what the value and purpose of such a discipline could be, given that it is not in revealing some absolute truth. The present work will offer its own answer to this question, but in regard to syntheorology alone, we are not making an inquiry into truth itself, but into that capacity which seeks, envisions, and desires it. As such, we are not interested in whether our theories are true, or in developing theories that are more likely true; we are interested in what it is possible to conceive of as being true, and the relations between these conceptions.

An analogy for this from the sciences that I find helpful is in the discovery of chemical elements based on the underlying structure of the periodic table, which represents a matrix of chemical properties. In 1869, chemist Dmitri Mendeleev published the first periodic table of elements, and this table represented the different known elements on an underlying matrix determined by the atomic masses of those elements. Prior to this, an element could be described only when we actually discovered a physical example of it in nature, but with this table of elements, Mendeleev was able to predict that there should be real elements which fill the gaps in the table. By studying the matrix of chemical elements, Mendeleev was able to accurately predict specific properties, such as the mass, density, melting point, solubility, and volatility, of then undiscovered elements such as gallium, which was discovered in 1875, and germanium, which was not discovered until 1996.

There is an obvious distinction between developing a description of a new element based on empirical analysis and developing a description of a new element based on its location upon an underlying matrix. Syntheorology is comparable with the latter, and so as we further understand the intricate details of the underlying matrix of concepts, we will be better able to categorise those concepts periodically based on their locations upon it. It is in this sense that I consider syntheorology to be a kind of archaeology of consciousness—an excavation of the field of potential belief, as opposed to the construction of those beliefs from the ground up.

In the present study, we have identified a basic quadripartite structure, but further analysis will surely identify subclasses within the four and may even reveal an inherent fractality to the matrix. Philosophical theories, like empiricism or materialism, can be regarded as points of resonance across the structure. Syntheorology affords us the ability of talking about these points, across disparate philosophical considerations, similarly to how we talk about shapes being regular, colours being complementary, or sounds being harmonic.

Just as the consonance and dissonance of musical chords can be explained mathematically by way of the geometrical relations of notes, syntheorology will be able to formalise what it is precisely that makes two distinct philosophical concepts go along together or not. If we are to consider thinking as an art, we should expect this, for we do not seek in musical theory, by analogy, to show that one chord is better than others, even though there are undoubtedly some that sound 'nicer' to our ears. We rather seek to understand how they all are related, what they are basically made of, and how they can be combined in the composition of a symphony.

A Morphology of Belief

The dialectical matrix will reveal that our ability to reason and the products of our reason are isomorphic. An isomorphism occurs when two structures can be mapped onto each other in such a way that each part of one structure corresponds to an equivalent part of the other, and where corresponding parts play similar roles within their respective systems. As such, it would be impossible to distinguish two isomorphic structures solely from the shape of the structure they share—in this case, the dialectical matrix.

This consideration leads us on to the third way in which syntheorology is a useful tool, which is as a morphology of a certain kind of belief system. Since the dialectical matrix is universal in regard to philosophy, it occurs in the same form within all of its individual branches. It matters not whether our inquiry concerns knowledge or being or virtue—all positive philosophical notions emerge from this uniform structure, and so the content of all topics will reveal themselves as mutually congruent, with each concept from one topic being isomorphic with those from others. There is thus an epistemological notion for each in ontology, and logic, and the philosophy of mind, and formal theories of truth, and the isomorphisms between them reveal the underlying form of our capacity to believe, as it permeates the lens of our means of experience.

In other words, one of the aims of syntheorology is to uncover precisely which conceptions, from distinct areas of discourse, correspond to the same points of resonance across the underlying matrix. In this way, syntheorology is an examination of the archetypal structures of consciousness as they constitute humankind's rationalisation and conception of its world. We have seen that the interaction of two dimensions of belief compose four basic classes of theory, defined by the oppositions of *immutability* and *mutability*, *determinateness* and *indeterminateness*. Concepts that share the same belief-satisfying properties are *syntheoretic* concepts since they are grounded in the same part of the matrix.

'Immutability' is syntheoretic with 'feeling' in Jungian typology, with 'analyticity' in epistemology, 'universality' in ontology, 'absoluteness' in axiology, and 'collectivity' in political theory. Immutability manifests in all of these concepts in their wholeness and singularity, for the components of a whole may change while the whole itself remains complete. Conversely, 'mutability' reveals itself in its fragmentation and plurality, as revealed in the concepts 'thinking', 'synthetic', 'particular', 'relative', and 'individual'.

For the other dimension, 'determinateness' is syntheoretic

with 'extraversion', 'aposteriority', 'concreteness', axiological 'subjectivity', and 'liberty'. These concepts are related in their reliance on objectivity, physicality, and their determinability through an external environment. Finally, 'indeterminateness' is syntheoretic with 'introversion' in Jungian typology, 'apriority' in epistemology, 'abstractness' in ontology, 'objectivity' in axiology, and 'authority' in political theory. These concepts are related in their reliance on subjectivity and epistemic idealism.

It is important to note here that the fact of two concepts being syntheoretic does not suggest that a person is irrational if they only believe in one of them, but rather that they rely on the same belief-satisfying property. Nevertheless, we may well argue that an idea gains a point in its favour wherever it correlates with something else we believe to be true. The theoretical classes arising from the interaction of the two dimensions of concepts can be grouped to form more general syntheoretic ideologies, and these ideologies express a novel form of parsimony in that they postulate just a single category of belief. For these ideologies we shall use the labels given in

	Indeterminate	Determinate
Mutable	**Abjectivism** Synthetic A Priori Abstract Particular Objective Relative Authoritarian Individual	**Objectivism** Synthetic A Posteriori Concrete Particular Subjective Relative Libertarian Individual
Immutable	**Subjectivism** Analytic A Priori Abstract Universal Objective Absolute Authoritarian Collective	**Superjectivism** Analytic A Posteriori Concrete Universal Subjective Absolute Libertarian Collective

Chapter Four for the Jungian temperaments: 'subjectivism', 'objectivism', 'abjectivism', and 'superjectivism'.

Trying as we can not to trample over travelled ground, we shall now proceed with a brief description of these four syntheoretic ideologies. We do this in the effort of showing that the concepts and theories within them are compatible and mutually implicative, as to further reinforce that the dialectical matrix is an effective method for systematising the dynamics of philosophical ideation.

Syntheoretic Ideologies

We begin with *subjectivism*, which is oriented by *immutability* as a belief-satisfying property, and an acceptance of *indeterminateness* as belief-disrupting. These are both passive and negative properties, and so subjectivism features a bias towards the passive and negative aspect of the theoretical duality. It thus relies on negative characterisations of phenomena and the coherence of ideas within a larger field of discourse. Its focus on relations conditions it to the underlying meanings of concepts, rather than the individual reasons *for* them, and these meanings are apprehensible only in terms of what a concept *is not*, thus implying an integral connection between the part and whole. This description is fitting for we will find that the subjectivist ideology is also syntheoretic with negativist approaches to epistemology in science, as characterised by the critical rationalism of Sir Karl Popper. In contrast to the externalist approach of the logical positivists, the syntheoretic subjectivist opts for logical *negativism*—for deduction, not induction; for falsification, rather than verification.

Psychologically, subjectivism manifests as *feeling*-based judgement and *introverted* attention, which is based on an intimate connection to values, qualities, and principles. In epistemology, it adheres to the *analytic a priori*, which concerns necessary truths, in virtue of meanings, that hold prior to induction and external conditions. In ontology, it is concerned with *abstract universals,* which are qualities that have a necessary connection with their instances; and in axiology, with the *objective absolute,* which reflects necessary values that hold universally. Finally, in political theory, subjectivism manifests as *authoritarian collectivism,* expressing an immutable state of egalitarianism and a restriction on the self-determination of individual will. As such, the paradigmatic syntheoretic subjectivist is an analytic rationalist or logicist, a subjective idealist, a moral absolutist, and an authoritarian socialist.

Objectivism is the direct contrary to subjectivism, being devoted to *determinateness* and accepting of *mutability*. These are active, positive properties, which also make objectivism a biased ideology, and I highlight that the term 'biased' is not used in a derogatory or negative sense here, but simply as the opposite of 'neutral' or 'centrist'. The objectivist thus relies on positive characterisations of phenomena and is fragmentalistic in its focus on discrete facts or reasons that are built from the ground up, placing less emphasis on how they fit together within a larger picture of reality. Such atomism of truth is characteristic of logical positivism, induction, and verificationism.

Psychologically, objectivism manifests as *thinking*-based judgement and *extroverted* attention, which form a collaborative kind of reasoning that focusses on the external. In epistemology, objectivism observes the *synthetic a posteriori*—to induction through empirical experience and external determination. In ontology, it is oriented by *concrete particulars,* which are individual spatiotemporal objects that interact with our instruments of observation. In axiology, it is concerned with the *subjective relative,* which reflects unreal values, dependent on the mind, applying relativistically to circumstance. Finally, in political theory, it manifests as *libertarian individualism,* which values the right to individual liberty, and is accepting of the consequent inequality. Thus, the paradigmatic syntheoretic objectivist is an empiricist, an eliminative materialist, a moral relativist, and a libertarian capitalist.

The third ideology is that of the *abjectivist*, which is located at the top left of the dialectical matrix. This is a combination of the belief-disrupting properties *mutability* and *indeterminateness*, and so the abjectivist is conditioned by an absence of the belief-satisfying properties *immutability* and *determinateness*. What satisfies the belief of the abjectivist is its own lack of bias, for while it may see these properties as desirable, it does not do so for one any more than another. It also recognises the contradiction between them, and so the irrationality of exalting them both becomes less attractive than accepting the unfixed and unobserved, which are wholly consistent, and in which the abjectivist finds the pinnacle of reason. Thus, while the abjectivist retains their levelheadedness, they sacrifice an understanding of the world based on their intimate connection to experience. They renounce the ideal and disavow the sensible, their realism carrying with it an absence of optimism, but permitting their pragmatic approach to knowledge.

Abjectivism is syntheoretic with *introverted thinking*, which is an inward-turning reason that arrives at its conclusions without influence from an outside source. In epistemology, this manifests as the *synthetic a priori*, which is acquired prior to experience, but also through intuition, in that a conclusion is not logically necessitated by its premises. On the Kantian analysis, this refers to mathematical, philosophical, and abstract scientific inquiries that are neither determinate nor immutable. In ontology, it is concerned with *abstract particulars*, which are particular instances of properties, or mathematical objects, such as numbers and propositions. In axiology, it is concerned with the *objective relative*, which refers to a plurality of conflicting real values, whose objectivity provides a guide to action rather than an absolute rule.

Finally, in political theory, abjectivism manifests as

authoritarian individualism, which most clearly expresses the abjectivist's need for equanimity and stability. Authoritarian individualism attempts to provide a middle ground between economic equality and social liberty, which necessitates a degree of inequality and authority, while allowing neither to grow disproportionately large. It recognises the moral worth of individual freedom, but also the importance of laws and restraints that would protect them and preserve cultural identity and tradition. Thus, the paradigmatic abjectivist is a synthetic rationalist; a trope theorist, mathematicist, or neutral monist; a moral pluralist; and an authoritarian capitalist.

The fourth and final syntheoretic ideology is *superjectivism,* which is dialectically opposed to abjectivism, being a combination of both belief-satisfying properties. Like the latter, superjectivism is neutral in that it does not express a bias in favour of immutability or determinateness respectively, but unlike it in that it does not take their equitability as a cause for concession. It takes faith in their union, rather than deconstruction, beyond the sphere of impressions, willing to risk foolishness if it preserves the harmony of their perspective. Nevertheless, the perspective is replete with contradiction, and this may be seen by others as pure naivety.

In Jungian typology, superjectivism is syntheoretic with the extraverted feeler, which Jung describes as being concerned with objective reality, but only insofar as they can reconcile it with abstract and eternal principles that arise from the unconscious. As such, far from forsaking the positions of those biased types, what they seek is a fusion of the subject and the object, which is done, if not from the position of some mystical insight, at least with the sincerest of intentions.

Epistemologically, it manifests as the *analytic a posteriori,* which refers to justification based conjointly on experience and necessity. Since analytic truths are knowable prior to any experience, it is the received view that the analytic a posteriori is a contradiction of terms. In ontology, it is concerned with the *concrete universal,* which is an entity that is simultaneously spatiotemporal and one which others share in common. Save for the Hegelians, the concrete universal is not generally taken to represent a genuine class of object, universals being ideal, and concreta fragmented. We might say that the concrete universal is the basic substance of the universe at large, though this is never an object of our direct perception.

In axiology, it conforms to the *subjective absolute,* the value of which depends on minds while also applying universally. Subjective absolutism is untenable as the foundation of a normative theory, since moral subjectivism implies that value is relative, but it is possible meta-ethically as *ideal observer theory*. Such a theory is nevertheless defined circularly insofar as valid value judgements and those who make them are defined in terms of each other. Finally, in political theory, superjectivism manifests as *libertarian collectivism,* which relies on the complementarity between equality and liberty, just as the superjectivist in general relies on the complementarity of immutability and determinateness. As discussed in the previous chapter, in civil life, we are often forced to make a choice between these two rights, for so long as individuals are not ideal moral agents, equality will require the state to limit the freedom to be unequal.

In conclusion, it is clear to see that applying the dialectical matrix to the various disciplines of philosophy reliably produces consistent worldviews. The polarity between syntheoretic subjectivism and objectivism expresses the standard dialectic between the mental and material, and adherence to one side of this divide qualifies these ideologies as biased or one-sided. Regarding the former, the introverted feeler, looking inwards at the structure of their mental states, naturally adheres to what is grasped of necessity and remains immutable, building their metaphysic from the ideal forms of their internal experience.

They have no trouble positing the mind-independence and absoluteness of value, finding no material conditions that could make them be relative, and therefore also good reason to condone their enforcement.

Regarding objectivism, the extroverted thinker is bound to the outside world, loyal to the determinate empirical process, and building their metaphysic of the concrete forms of corporeal reality. More theoretical areas of physics pose a challenge for the objectivist, for they see no reason to recognise the existence of entities that are not strictly sensible. While the abjectivist would have no issue maintaining confidence in, for example, the existence of quarks as a necessity of mathematical models, the objectivist requires its proof to come directly from its domain. Such proof is unavailable for any non-natural values, nor for anything non-empirical, and insofar as they remain immune to the methods of physical science, there can be no justification in enforcing them on others.

The remaining ideologies—abjectivism and superjectivism—are akin by their neutrality just as the former are by their biases, and just as these reach out in opposing directions, so too does this neutrality find expression in opposing forms. These forms are defined by the distinction between contradiction and complementarity, for while the abjectivist sees unity in the transcendence of opposites, the superjectivist finds it immanent within them. That is to say, the abjectivist sees the contradiction of opposites as a reason to look past them—to forsake them as artefacts of experience, which are not fundamental. It is a minimal expression of neutrality for its two cups are empty, and they must see things this way for anything else is either a violation of their logic, or a violation of their equity. The superjectivist, on the other hand, expresses maximal neutrality—its cups equally filled—for they are loyal to their perceptions, and take faith that the opposites are in fact complementary. Thus far, we have characterised the superjectivist in much a traditional way, as the exponent of paradox and fallaciousness. However, once we solidify the rationale of complementarity within the nature of truth, we can begin to see them in a new light, and on this we have much more left to say.

Part III

On the Duality of Logic

Chapter Twelve

The Laws of Thought

Introduction

The ideologies developed in Part II are formed of the combinations of dialectical properties. In accordance with a monoletheic and monistic approach to philosophy, which can only hold one property of a contradictory pair to exist in the same thing at the same time, if any one of these ideologies reflect reality, the rest of them cannot. While we can accept the validity of both a priori and a posteriori justification, we cannot posit a reality made both of matter and mind, where matter is concrete, and mind is abstract. We can claim, as the abjectivist might be inclined to do, that both are expressions of a neutral substance, abstract and yet particularised, but this view would contradict that of the objectivist in its abstractness, the subjectivist in its particularity, and the superjectivist on account of both.

For the monoletheist, there is a way in which the world is like, and amongst contradictory views, all but one must be delusions. Monoletheism is a natural part of our experience of the world, just as it is in our discourse, and when we are met with someone holding an opposing view, it is natural that we should become competitors with them in the search for truth. The more we admit to the rationality of our foes, the more we discredit our own positions, and so we must feel strongly, challenge harshly, and take faith that we *can* have opinions that converge on how things are.

But now, we are presented with a curious realisation: our views are not chaotic, as monoletheism might suppose, for they conform to a predictable structure. We have not a continuous range of views to be whittled down from the outside in, as we do in science, but a clear dialectic that is unwilling to compromise.

We must understand what, if anything, this means. Is it that we are conditioned to this same dialectic by virtue of the mode of conscious experience? Are we not well suited for philosophy? How can we be sure that our beliefs are not artefacts of how our organism works, advancing matters utterly disinterested with truth? And how can we be sure that our reason ever touches the real world?

We have a distinctly Kantian problem, arrived at in a new way. Syntheorology reveals to us that there are *ways* we conceive the world as like, and while there is variation within these ways, the ways themselves are totally firm. The structure permeates not only our ability to conceive, but what we believe psychologically, and how we perceive biologically. Either, we are perpetually bound by the structure of our being, never able to see the real world, or that which binds us is the world itself. We are at a similar crossroads as that which separated Kant and Hegel two centuries ago. It is the latter option that would make monoletheism false, while in the former it is simply unknowable.

In this second part of the present work, we shall approach the concept of truth from the perspective of formal logic, for logic is the one area of philosophy that should supersede syntheorology. We cannot disagree on what it *means* to be true, can we? Well, as we shall see, truth itself is a notion just like 'being' or 'value' that can be characterised by the dialectical matrix, and this fact will provide us with the key to making progress in philosophy. To arrive there, we shall have to be relatively comprehensive in our approach, so that we do not miss the significance of the conclusion. I realise that the following three chapters will be somewhat of a history lesson for the logicians among my readers, and a little intensive for those who are not, but it is important to be cognisant of the development of our thought in this area as context for later on.

In the remainder of the present chapter, we shall discuss Aristotle's syllogistic logic and its later formalisation by George

Boole, with particular focus on the laws of thought. In Chapter Thirteen, we shall discuss Frege's predicate logic and the paradox it was found to entail; and in Chapter Fourteen, we'll discuss how we might avoid the paradox as well as Gödel's incompleteness theorems, with some focus on the Hilbert program.

Syllogistic Logic

There are few ideas that remain as constant in philosophy as those that are found in the field of logic, and it was not until around 200 years ago that we found the need to alter our methods significantly from how they were formed in ancient Greece. While appeals to logical fallacies predate logic as an independent subject of study, it was Aristotle who formalised our common-sense notions into the first logical system. Aristotle took our conceptions of what exists to be bound by laws governing their behaviour, and for Aristotle, there was one law which ruled above the others—the *law of non-contradiction*. This law reigns supreme because it is required for rational thought to take place. In Book IV of the *Metaphysics*, he describes this "most fundamental principle of knowledge" as a law which is impossible to mistake, being one which all people who understand anything must know by default. He writes:

> He whose subject is being *qua* being must be able to state the most certain principles of all things. This is the philosopher, and the most certain principle of all is that regarding which it is impossible to be mistaken; for such a principle must be both the best known (for all men may be mistaken about things which they do not know), and non-hypothetical. For a principle which every one must have who knows anything about being, is not a hypothesis; and that which every one must know who knows anything, he must already have when he comes to a special study. Evidently then such a

> principle is the most certain of all; which principle this is, we proceed to say. It is, that the same attribute cannot at the same time belong and not belong to the same subject in the same respect.[119]

Here, Aristotle presents the law as a modal claim on the nature of objects, and he goes on to present it as a psychological claim on the nature of belief. Most importantly, however, it is a semantic claim on the nature of truth, in which form it states simply that "contradictory statements are not at the same time true".[120]

The law applies universally and precludes any proof of itself, for there are no prior axioms from which it could be deduced. To deny the law of non-contradiction is to be unable to say anything definite or meaningful, for language itself is predicated on its veracity, as is our ability for intention. Moreover, to deny the law of non-contradiction is to deny that an object has any essence or substance at all, since, for Aristotle, substances are immutable and can never assume contrary states. Implicit in the law therefore is another, which holds a similar status at the foundations of rational thought. The law of *identity*, Aristotle says, asserts "that the word 'be' or 'not be' has a definite meaning, so that not everything will be 'so and not so'".[121] In other words, every word has a definition, for if all words meant all things, they would equally mean nothing at all, and so communication using language would be impossible.

Finally, the law of *excluded middle* completes Aristotle's triad of laws. As he states, "There cannot be an intermediate between contradictories, but of one subject we must either affirm or deny any one predicate,"[122] by which he means, in regard to a proposition and its negation, one of them must be true. This law presents the complementary case to that of non-contradiction, for while the latter asserts that contradictory propositions cannot both be true, the law of excluded middle states that both cannot be false.

These three laws appear to be unquestionable for they are prerequisite of all semantics, our language being not merely a tool for conversation, but embedded into the structure of our thought. Our words carry meaning insofar as we give meaning to them, and the laws of thought are the glue which binds these meanings to our capacity as rational beings. Taken together, the laws also appear so related that Arthur Schopenhauer once proposed to reduce them to a single axiom, stating, "Every predicate can be either confirmed or denied of every subject."[123]

Aristotle's system of logic, of which these laws form the foundation, proceeds by way of syllogism, and a syllogism is simply an argument in which a conclusion is deduced from two or more premises. For example:

All men are mortal.
Socrates is a man.
Therefore, Socrates is mortal.

Aristotle explains that there are four basic types of propositions within syllogistic logic, which are formed by two dichotomies in much the same manner that we have described for the dialectical matrix. The first dichotomy, between 'universal' and 'particular' propositions, dictates whether the predicate concept applies to all or just some of the subject, and the second, between 'positive' and 'negative' propositions, dictates whether the predicate concept is affirmed or denied of the subject. Thus, there are four classes of propositions: the universal affirmation, such as 'all men are mortal'; the universal negation, such as 'no men are mortal'; the particular affirmation, 'some men are mortal'; and the particular negation, 'some men are not mortal'. These four types of statements are referred to as 'categorical' because they designate whether the members of one category are members of another, and this is important because it is the earliest allusion we have to the notion of sets, which would become instrumental in the development of later logics. The propositions can in fact be depicted as sets, as they were by John Venn in the late 19th century.[124]

An important point of note at this point in our inquiry is that Aristotle's logic is distinctly metaphysical in nature. He claims that everything that exists can be formulated into a proposition of the form 'subject copula predicate', which asserts or denies its containment in a category. A thing is made true by its relation to a substance, which is to say, a statement like 'all widows are women' is true because the property 'womanhood' is a part of the substance 'widow'. It is what the terms of the logic refer to in the world that determines their truth, and not solely the arrangement of those terms, distinguishing syllogism from mathematics, which is entirely about the arrangement of terms. There is nothing in the logic itself that dictates whether a predicate should be affirmed of a subject, and so Aristotle's logic is *bound* to the natural world and *depends* on objective experience.

Symbolic Logic

Aristotle's system remained essentially unchanged for well over two thousand years, and Kant once claimed that it contained all of logic that there was to know. In the mid-19th century, however, an English mathematician by the name of George Boole aimed to improve on Aristotle's work, by disentangling logic from the terms and categories that are dependent metaphysically on the world. Boole's logic would rest solely on numbers, equations, and symbols, and by 1854, he had successfully developed a series of algebraic rules, which could make self-evident deductions purely in terms of the *logical* objects to which our concepts refer.[125]

Unlike Aristotle's logic, which was intensive and relied on our comprehension of concepts, Boole's logic was *extensive* in that a concept now referred simply to the class, or set, which contains its instances. Thus, a proposition like 'all widows are women' could be expressed by saying that the class of all widows is contained within the class of all women, and so the objects of the logic are no longer substances embodying concepts, but purely abstract collections denoted by symbols.

Boole's system made logic an exercise of mathematics, using algebraic notation to express its terms, signs, and symbols. Letters represented different objects, names, and classes, and other symbols represented the logical operations by which they are "combined or resolved so as to form new conceptions".[126] Boole was then able to construct propositions of ordinary language purely algebraically. For example, if we let x = 'men' and y = 'married people', then xy is equivalent to 'men who are married'. The conjunction of the class of all men and the class of all married people is denoted by multiplication, $X \times Y$, and represents the class of all men who are married. Similarly, the disjunction of the class of men and the class of married people is denoted by addition, $X+Y$, and refers to the class of all men or married people. The inverse operation of subtraction, $X-Y$, then refers to the class of all men who are not married.

The law of identity can now be written algebraically as $xx=x$, meaning 'men that are men equals men', and in order to derive the laws of non-contradiction and excluded middle, Boole introduces two special classes in terms of which he can define all others. The first of these special classes was called the 'universe of discourse' and is related to what we would now call the 'universal set'—the set of all those things we are talking about within a given proposition. That is, the totality of what it is we are considering in a logical inquiry contains all contrary notions with respect to some 'universe', which is the "field within which all the objects of our discourse are found".[127] If we are considering the properties of human beings, then the limit of our discourse is the set of all humans. This universe contains the contraries 'men' and 'women', or 'married' and 'unmarried', but there are no non-humans within it. The second special class was called 'nothing', mirroring the more modern 'empty set', and refers to a field of discourse in which there are no objects.

To express these special classes algebraically, Boole had to give them also a symbol, and the only symbols that fit for Boole were the integers 1 and 0. This is because 1 and 0 are the only numbers that satisfy the law of identity, $xx=x$, when substituted for x. Individual objects and classes are then defined in relation to these classes. For example, if the universal class 1 is the class of all humans, and is the class of all men, then $1-x$ represents the class of all humans who are *not* men.

From these foundations, Boole was able to derive the laws of thought purely in the language of symbols. The law of non-contradiction asserts that the intersection of a class with its negation contains no members and is thus expressed $x(1-x)=0$. If we substitute in the terms of our previous example, we get 'the class of all men who are not men is the empty class'. Furthermore, the law of excluded middle asserts that the union of a class with its negation contains everything (in the universe of discourse) and is expressed $x+(1-x)=1$, or 'the class of all men or not men is the universal class'.

Boole's algebra provided a method of employing Aristotle's syllogistic logic in terms of computational procedures, wherein the laws of thought are formulated in terms of membership to classes. His work laid the foundations of algebraic logic, and over the following decades his system was refined into what we now refer to as 'Boolean algebra'. The operations of Boolean algebra were also given diagrammatic representation as sets by John Venn, and in 1936, it was shown that Boolean algebras in general are isomorphic with fields of sets.[128] Boolean operations with regard to different variables can thus be represented by the intersection and union of sets, which correspond to the logical connectives 'conjunction' and 'disjunction', and the natural language terms 'and' and 'or', respectively. The architecture of computing, and by extension the inauguration of the age of information, was also facilitated by Boole's work, as binary programming languages are simply physical implementations of Boolean algebra.

Modern Boolean algebra has the values 0 and 1 as meaning *false* and *true* respectively, which reflects their usage in electronic circuitry, with *open* and *closed* logic gates, and the states of 'off' and 'on'. However, the law of non-contradiction, $x(1-x)=0$, does not state that the conjunction of a proposition with its negation is false, but rather that it is a contradiction. Similarly, the law of excluded middle, $x+(1-x)=1$, does not state that the disjunction of a proposition with its negation is true, but rather that it is complete. Boole's implementation of the values 0 and 1 in logic reveals our early inclination that contradiction should be associated with falsity, and that completeness should be with truth. As logicians William and Martha Kneale noted in *The Development of Logic*, "We have all that is needed for an interpretation of Boole's system in terms of the truth values of propositions with the symbols 1 and 0 standing respectively for truth and falsity."[129]

The significance of having just two values—true and false, on and off, good and bad—is that so long as we maintain that opposites contradict each other, which is itself the basis of our efforts to bifurcate the universe of discourse, it is always possible that we can arrive at a definite outcome. Recognition of Boolean logic within philosophy makes it appear essential that one side of a dichotomy should enjoy an absolute share of the truth, and we have built into the structure of our language the demarcation of the right from the wrong.

If one side of a logical duality is held to be right and just, and if this logic is the basis of rational thought, then it is natural that we should seek some correspondence to halves of other dualities from our experience as well. The correspondence theory of truth is as old as philosophy itself and asserts that a proposition is true precisely when it corresponds to something in the world, and so, equally, false beliefs are those that do not refer to anything. We accordingly assume that duality is just a convention—a consequence of the limitations of perception—and that there is no true duality at all. We see that truth is absolute, and that all else is delusion about the non-existent.

If we are to hold that it is possible to be right in any thesis, or at least to encroach upon what is right, then we cannot be blamed for wanting to fall on the right side of the divide. To defy the principles of logic intently is never favoured over making mistakes in the name of progress, and so we find it rational to pledge allegiance fully to our inclinations, and disparage the views of others. Just as we construct the rules of logic from the only way intelligible thought can be, we select our beliefs in terms of what we most readily see. If we must pick a side, then the overt and obvious is our safest choice. Yet, it is in the nature of opposition, whether contradictory or complementary, to express the overt *in relation* to the covert, and so what seems an obvious choice on the surface hides within it an integral relation, that cannot be removed without reducing what attracted us to our beliefs in the first place. This relation is itself the foundation of our logic, and so our opposites *must* go together.

Chapter Thirteen

The Paradox of Self-Reference

The Universal Characteristic

Boole's logic taught us that it is not merely the content of what we say that determines its truth, but also the logical form in which our claims are made. Aristotle's syllogistic logic was metaphysical, insofar as a proposition is true simply when it affirms an existent in the world, but the transition to a logic of symbols allowed us to disentangle our arguments from their referents and judge them purely in terms of the relations of arithmetical objects such as classes and operators.

Nevertheless, the operators of modern symbolic logic are all contained in natural language—the conjunction (∧) symbolises the word 'and'; the disjunction (∨), 'or'; and the negation (¬), 'not'—and so its significance was not that it is independent of our intuitive understanding of language, but that it allows us to manipulate the symbols without having to *appeal* to what they represent in language. Thus, while the rules of inference are devoid of metaphysical relations, the symbols of the logic still do depend on an understanding of what those symbols represent. As soon as we want to be philosophical about things, and extract meaning from our logic, we are required to refer to what the symbols mean metaphysically, and this process is fundamentally reliant on the use of human intuition.

It is natural that we should ask whether it is possible that we might construct a language whose symbols are bonded to our concepts *logically*, and not merely because we all agree on them. It was this consideration that caught the mind of the 17th century German rationalist and polymath Gottfried Leibniz, whose inquiries into formal logic remained overlooked in great part until the 20th century. As Bertrand Russell surmised,

Leibniz "did work on mathematical logic which would have been enormously important if he had published it; he would, in that case, have been the founder of mathematical logic, which would have become known a century and a half sooner than it did in fact".[130] As Leibniz did not publish his work on the subject, the founding honour belongs to George Boole. However, Leibniz's intentions for logic feature an important distinction from that put forth by Boole, for he was particularly sensitive to the importance of the relation between language and logic for philosophy.

Leibniz recognised ordinary language as merely a tool we have invented for communication, and one that is not particularly well suited for our inquiries in philosophy or logic. He envisioned that we might create a *characteristica universalis*—a universal language—whose symbols directly resemble the concepts of science, mathematics, and philosophy. Such a symbolic logic would contain the axioms of knowledge as its foundation and would allow for new information to be generated in a systematic way. "We should be able to reason in metaphysics and morals in much the same way as in geometry and analysis," he writes.[131] The characters of such a language would be ideographic, like Egyptian hieroglyphics or many of the scripts of Asia, and their meanings would be natural and not merely conventional. The structure of the language would then represent the relationships between ideas and would be combined with a system of algorithms—a *calculus ratiocinator*—which would manipulate the symbols and calculate whether strings expressing philosophical ideas could be produced.

Leibniz was, of course, aware of the magnitude of such a task, but he sincerely believed it to be possible in principle. As Russell notes, he made steps towards his goal which were recreated and furthered by Boole and his successors, but the reliance on our intuitions of the relations between symbols and their meanings, in 19th century logic, meant that we were still a long way away from representing all philosophical propositions in a purely logical form.

Leibniz's vision was left dormant for over two centuries, but in 1879, a radical development in logic was set into motion with the publication of German philosopher-logician Gottlob Frege's *Begriffsschrift*, or 'concept-script'. This landmark work would mark the transition between the symbolic logic of Boole, and a new formal logic, which aligned much more closely with Leibniz's conception of a universal language, built not on convention, but on logic alone. Frege's work would also mark a true departure from the metaphysical commitments of previous logics—a goal that was directly inspired by Leibniz.

The Concept Script

Frege began his efforts from the perspective that Boolean logic was limited in that it served as a tool for drawing conclusions from premises but was insufficient for determining which premises should be admitted in the first place. Frege wanted to construct a formal language using the symbols of logic which extended *into* mathematics, so there is a major difference between the ways in which Boole and Frege viewed the relation between the two. While Boole treated both mathematics and logic as emergent from an understanding of the "general principles founded in the very nature of language",[132] Frege was motivated by a belief that the epistemic status of arithmetical truths was purely logical and analytic, not intuitive and synthetic as Immanuel Kant had supposed.

This rationale was the basis for a position and program in the philosophy of mathematics called 'logicism', which aimed to reveal that mathematics was purely an extension of logic. Logicism begins from the traditional laws of thought, enunciated by Aristotle and which are justified of necessity, and attempts to derive the whole of mathematics from these primitive foundations. As Frege describes, his task was "to ascertain how far one could proceed in arithmetic by means of inferences alone, with the sole support of the laws of thought".[133]

In revealing mathematics as a purely logical exercise, Frege would eliminate the need for synthetic reasoning and human intuition, thereby providing the basis for a genuine calculus ratiocinator. We can imagine, today, such a system of algorithms working as programs upon the framework of modern computing, being able to calculate philosophical, mathematical, and scientific truths mechanically from the logical structure of the ideas that form them. The logicist program was also important in that it was the impetus behind the tradition of analytic philosophy, which sought to reduce the world of reason into its elementary constituents, using formal logic as its guide and arbiter. If logicism was to be successful, not just for mathematics but for philosophy as a whole, it would then be possible that any true theory of philosophy would be a theory derivable from the principles of logic.

In the introduction to *The Foundations of Arithmetic,* Frege states his guiding principles: "always to separate sharply the psychological from the logical, the subjective from the objective; never to ask for the meaning of a word in isolation, but only in the context of a proposition, and to never lose sight of the distinction between concept and object".[134] This distinction between concepts and objects again derives from Frege's analytic approach to the relation of logic and language—an important departure from earlier logics. Ernst Schröder, who helped to develop the modern form of Boolean algebra, described human understanding as being founded upon a psychological process of association between names and objects. This is a distinctly synthetic process for names constitute natural language and objects constitute the natural environment. In order for communication to work we must psychologically stitch the two together in a collective agreement that words refer to respective things.

In Boole's logic, a proposition is merely a union of two terms, expressed by the 'subject copula predicate' structure; these propositions are the basic elements of the logic—they cannot be broken down into more useful parts. Frege explained that a single idea must signify *both* a concept and an object, such that a proposition is precisely a concept that is applied, or predicated, to an object.[135] Since concepts can be asserted of anything, that thing may be an object, but it could just as well be another concept, which we should be able to extrapolate into its own 'subject copula predicate' proposition. This more considerate analysis of the structure of propositions only works if we have a way to distinguish their respective components—something previous logics were simply unable to do.

Frege also introduced a way to *quantify* over objects and concepts, further expanding the kind of propositions we are able to model. The declarative 'subject copula predicate' sentences of Boolean propositional logic are incapable of capturing variables that can be instantiated by multiple subjects, and so many intuitive arguments of natural language simply cannot be expressed in Boole's logic. Propositional logic is a way to represent propositions with symbols, but it cannot represent objects, and collections of objects, within those propositions. Consider the following syllogism of natural language:

a. All men are mortal.
b. Socrates is a man.
c. Socrates is mortal.

Each of these propositions is symbolised in propositional logic, and so the entire argument is expressed as $a+b\Rightarrow c$, which is invalid as there is no logical connection between the individual declarative sentences. In a sense, propositional logic cannot 'see' the parts of the sentences that validate the argument, it can only see the sentences as whole units. What we need is a way to assign properties to the elements of the sentences so that we can then reason about these properties, and in Frege's *predicate* logic this is done via quantifiers and variables. The argument thus becomes:

1. For any x, if x 'is a man', then x 'is mortal'.
   $\forall x(man(x) \Rightarrow mortal(x))$
2. Socrates 'is a man'.
   $man(Socrates)$
3. Therefore, Socrates 'is mortal'.
   $mortal(Socrates)$

In this example, x is the variable, and 'for any' expresses the quantifier, here the universal quantifier $\forall$, which asserts that a given property holds for every member in the domain of discourse. The atomic sentence has now been broken down into a compound statement and shares a common term with the second premise and conclusion, making Frege's logic powerful enough to represent complex arguments purely in terms of logical notation.

We now have all we need for a theory of concepts and objects in terms of membership to sets, where concepts contain and predicate over other concepts. To this end, Frege defines a particular kind of object called an 'extension', which refers to all of the objects over which a particular concept extends—essentially, a set. For instance, for a concept 'red', the 'extension of red' is the collection of all those objects to which 'red' is applicable. Extension is expressed formally by way of the most famous of the axioms of Frege's calculus, *Basic Law V*, which in essence asserts that for any concept there is a set of objects over which that concept extends.[136] In accordance with this law, the meaning of a logical object is no longer dependent on the meanings of words but is simply the set to which it applies.

Frege goes on to use extension to provide a definition of the natural numbers, which are defined by the quantity of objects falling under concepts. Thus, the number 0 is the extension of the concept of all concepts under which nothing falls, which for him is the concept 'being non-self-identical'. The number 1 is then the extension of the concept of all concepts under which one object falls, which Frege says is the concept 'being identical with zero', since only one object falls under this concept, that object being 'zero'.[137] The number 2 is defined as the number of the extension of all concepts having two members, such as the concept 'moons of mars', since mars has two moons, and so on. Since many concepts will extend to precisely two objects, the *extension of the concept* of all those concepts is defined by an equivalence class relating these concepts together.

In defining natural numbers as properties of concepts rather than as properties of words or material things, mathematics is divorced from any intuitive notions of what it means to have two or three or four of something. The fundamental objects of logic are simply the extensions of concepts—sets—which have respective cardinalities, and this allows Frege to ground arithmetic within logic, wherein the concept of number is a direct result of the laws of thought.

In *The Foundations of Arithmetic,* Frege successfully showed that the Peano axioms, in which arithmetic is grounded, could indeed be obtained from his second-order logic. He began with a simple set of logical axioms, embodying the laws of thought, and proved that the whole of arithmetic could be formalised on this foundation. In doing this, he also seemingly proved that arithmetic is a purely analytic exercise requiring no human intuition, as purported by his philosophy of logicism. At the turn of the 20th century, therefore, it seemed as though Frege had done the impossible; he had achieved Leibniz's vision of a calculus ratiocinator *at least* in regard to mathematics, and he opened the door to the possibility of doing the same for philosophy and science in the future.

Russell's Paradox

Frege's proof that his system of second-order logic can produce the Peano axioms of arithmetic is referred to as 'Frege's theorem', and the key axiom in the proof is something now known as 'Hume's Principle'. This principle asserts that 'the number of *F*s is equal to the number of *G*s just in case there is a one-to-one correspondence between the *F*s and *G*s'. The principle is essential to Frege's proof for it is what allows him to define the natural numbers as the equivalence classes of all sets containing that number of elements. However, Hume's Principle is not an analytic law of logic but rather a product of human intuition, and so in order to utilise it in his logicist system he must derive it from some other, more fundamental law, which is his Basic Law V.[138]

Basic Law V in its modern form is referred to as the 'axiom schema of unrestricted comprehension', because it places no restrictions, in relation to natural language, on what predicates or concepts can have corresponding sets. It states that 'there exists a set *F* whose members are precisely those objects that satisfy the predicate *x*'. Thus, for *all* properties of objects there is a set containing those objects that have this property. There is nothing immediately suspicious about this principle; in fact, it appears to be self-evident, for what could make a set a set if not some property objects may share in common. Frege thus took it as a law of pure logic. Nevertheless, Basic Law V is not a direct consequence of the traditional laws of thought, and though it appears to be necessary, its use is where things fell apart for Frege. In 1902, while Frege was getting ready to publish the second volume of his *Basic Laws of Arithmetic*, he was sent a letter from Bertrand Russell conveying that a contradiction could be derived from his system, specifically from Basic Law V.[139]

Now, sets are collections of objects, but they are also objects themselves and so can be members of other sets. As we have discussed, each set is defined by a predicate, and the universal set is simply the set whose members are those that satisfy the predicate 'being a set', which means that the universal set must also contain itself as a member. 'Self-containment' is therefore a definable property of sets, and so we can also define a set whose members are those that satisfy the predicate 'containing oneself'; we can call these 'self-inclusive sets'. Conversely, sets that do not satisfy this predicate, we can call 'self-exclusive sets'.

We might be inclined to think that most sets will be self-exclusive, but there is actually a duality between these two types of sets via the negation of predicates. That is, the set of all cars is self-exclusive, but the set of all objects that are not cars is self-inclusive, for such a set is not a car. 'Self-inclusivity' and 'self-exclusivity' are inverse functions, and as we know from the laws of thought, every set must satisfy one or the other. The law of non-contradiction ensures that a set cannot be both, and the law of excluded middle ensures that a set cannot be neither.

Accordingly, we can define a predicate of 'being a self-exclusive set', and we can call the set which satisfies this predicate '*E*' for simplicity. Russell's paradox is revealed when we try to answer the question as to whether *E* satisfies the predicate 'being a self-exclusive set' or the predicate 'being a self-inclusive set'. In other words, is *E* a member of itself or not? Well, if *E* is self-exclusive, then it does not contain itself as a member, but given that *E* is the set that contains all self-exclusive sets, it is precisely the sort of set that itself should contain, and so it is self-*inclusive*. However, if *E* is a self-inclusive set, then it contains itself as a member, but *E* only contains sets that do not contain themselves as members, and so *E* is not the kind of set that itself should contain. Here we have the contradiction deduced by Russell; if *E* is self-exclusive then it is self-inclusive, and if it is self-inclusive, then it is self-exclusive. The result is unmistakable; if E is a set, it must contain or not contain itself, but it cannot contain or not contain itself, so E is not a set.

Frege's system is based on the fact that, in accordance with his Basic Law V, *all* concepts and definable properties have sets, yet the predicate 'being a set that does not contain itself as a member' cannot be asserted of objects and does not have a corresponding set, so *not all* predicates have sets. Russell's paradox is the ultimate example of a self-referential paradox demonstrated purely in the language of mathematical logic, and it is devastating for Frege's system because, from a single contradiction, we can infer any statement as true via the principle of explosion. In short, Frege's system, and the great achievement it carried with it, was mistaken.

Upon learning of the paradox, Frege was forced to write an appendix to his forthcoming second volume of *The Basic Laws of Arithmetic*. It begins:

> Hardly anything more unwelcome can befall a scientific writer than to have one of the foundations of his edifice shaken after the work is finished. This was the position I was placed in by a letter of Mr. Bertrand Russell, just when the printing of this volume was nearing its completion.[140]

In the wake of this tragedy, one might hope that the system could be tweaked in some way that the paradox could be avoided. Since Frege's theorem only requires Hume's Principle, one option was to simply drop Basic Law V entirely. However, Frege acknowledged that Hume's Principle was not a law of pure logic and so could not play a fundamental role in any system of logicism. Nevertheless, his system would indeed avoid the paradox while remaining largely unchanged if Basic Law V *was* replaced by Hume's Principle. For Frege, however, this was simply unacceptable, and this is because he had a much wider vision of the philosophical importance of his Basic Law V, which was for him irreplaceable.

While Frege was working at the field of mathematical logic, he was a philosopher at heart, and he was fundamentally motivated by his beliefs concerning the nature of logical objects. First and foremost, Frege recognised numbers as abstract particulars existing independently of our intuitions of material reality, and so they must be apprehensible by pure logic, without reference to the quantity of some other objects; this is something Hume's Principle cannot provide. Furthermore, if we cannot define numbers as distinct from other kinds of objects, such as material ones, we cannot demonstrate that any given object is a number or not. As Frege expressed, that certain objects are equinumerous does not "decide for us whether Julius Caesar is the same as the number zero". That is, Hume's Principle is only informative when the object presented to it is of the form 'the number of *F*s', and so it cannot provide the answer 'Julius Caesar *is not* the same as the number zero'. "What we lack is a concept of number; for if we had that, then we could lay it down that, if *q* is not a number, our proposition is to be denied".[141]

While Basic Law V is useful mathematically largely as the parent of Hume's Principle, it is the philosophical significance of the law that makes it so important for Frege and his larger logicist project. The fact that Frege is unwilling to compromise on his achievements—that he would rather take complete failure than partial success—tells us a lot about how Frege saw the basic purpose of his task. Sircun Gao summarises the requirement Frege set for himself: "*The logical foundation of arithmetic should be at the same time ontologically and epistemologically informative.* It is unacceptable even if arithmetic has only been *technically* reduced to logic."[142]

This is by no means stubbornness on Frege's part; the discovery of the inconsistency of his system not only destroyed Frege's academic productivity, but also to a considerable degree his mental health. He never published the planned third volume of *The Basic Laws of Arithmetic*—he did not work again at mathematical logic—and by the end of his life he expressed that he was no longer confident that mathematics is a product of logic.

Chapter Fourteen

The Incompleteness of Consistency

Type Theory

Naive set theory had been found to fail on the count of consistency, and Frege had withdrawn himself from his discipline. However, not everyone was ready to give up on the logicist program, and the torch was passed to Bertrand Russell, who set out to resolve the paradox. Russell recognised that if one eliminated our ability to define the universal set, then there could be no such object as 'the set of sets not containing themselves as members', and thus no paradox. This of course was not a simple matter, for universal sets, or at least universal collections, have been crucial for logic since the time of Boole. They are also highly intuitive, for if we can group together certain objects within a class, then it is natural that we should be able to group all objects together. Any visual depiction of a system of sets *necessarily* features some universe, if not explicitly then at least as the page on which one writes. There is always a limit to the objects being observed, and that limit was thought to be definable just like any other collection.

Russell wanted to preserve our intuitions about what kind of collections are possible, while also preventing any collection from extending over itself. He resolved to do this by arranging propositions into a hierarchy composed of collections existing on different levels. Russell referred to these collections as 'types', avoiding the now naive notion of sets, and types can only contain objects existing on lower levels of the hierarchy, and therefore never themselves. The lowest type consists of individual objects, while properties of individuals would form the next type higher, and since the universe of discourse has been stratified, an object can never be equated with the properties it might have.[143]

In order to justify the stratification of types, Russell would have to explain what logical reason prevents an object from extending over itself, and the unrestricted comprehension of classes, or sets, which is Frege's Basic Law V. The approach Russell takes is to introduce a 'vicious circle principle', which states that a collection cannot be a part of itself, and that the scope of a concept cannot extend over anything presupposed by that same concept. Why? Because, if it could, we would create the paradox.

Now, Russell recognised the strength of this principle in banning certain kinds of statements, and statements we intuitively should not want to ban. He notes that it prevents us from forming any proposition that refers to propositions in general; for instance, the proposition 'all propositions are either true or false' is a meaningless string of words in Russell's logic, it being itself a proposition referring to propositions. Despite this, it seems necessary that such a statement means something that is true, since it is the very essence of the laws of thought that propositions are either true or false. Furthermore, most instances of self-reference or circularity, like this one, are not vicious at all, and the principle only needs to eliminate those that are.

Here, we can begin to see the limitations of stratification, as it seems to imply that the logic of mathematics is *not* the logic of natural language. Yet, the laws of thought are topic neutral, meaning that they hold regardless of the medium in which they are expressed, and regardless of the subject-matter they are used to talk about. It is a useful analogy therefore to consider what stratification would look like if it was a part of ordinary speech. To borrow an example from Douglas Hofstadter, consider the sentence: "In this book, I criticize the theory of types".[144] Is this sentence meaningful? And is it true? Of course, it is on both counts; but if type theory were applied to natural language, then the sentence would be grammatically fallacious for this book cannot refer to itself. Hofstadter also points out that the word 'I' makes no sense in the context of type theory, for it has self-reference built into its own definition. We shall have more to say on this when we get around to Hofstadter's work later on in Chapter Seventeen.

Type theory characterises itself as being grounded purely in logic, yet it confronts the intuitions we have of what gives our assertions sense and meaning. As Hofstadter puts it, "The remedy [type theory] adopts for paradoxes—total banishment of self-reference in any form—is a real case of overkill, branding many perfectly good constructions as meaningless."[145] Of course, Russell didn't want to ban self-reference in natural language; he only wanted to do it in mathematical logic because he saw no other way out—we cannot have vicious circles in mathematics, and having no circles at all is one way to achieve this.

Paradox Revisited

The deeper problem to Russell's solution is that self-referential paradoxes are not just things that occur in mathematical logic, but in natural language too. Since many of these paradoxes have the same basic structure as Russell's paradox, it is natural to think that they should also share a solution—a sentiment that is encapsulated in Graham Priest's 'principle of uniform solution'.[146] If this is indeed the case, and if type theory *is* the solution to Russell's Paradox, then type theory is right for natural language too, and 'In this book, I criticize type theory' is grammatically incoherent.

To illustrate Russell's paradox in terms of natural language, Kurt Grelling and Leonard Nelson developed a semantic analogy, using words in place of sets. Just as Russell's paradox distinguishes between two types of sets—*self-inclusive* and *self-exclusive*—so Grelling's paradox distinguishes between two types of words—those that are *self-descriptive* and *non-self-*

descriptive respectively.[147] A self-descriptive word is one whose meaning applies to itself, such as 'English' or 'polysyllabic', for 'English' is an English word and 'polysyllabic' is a word of multiple syllables. A non-self-descriptive word, on the other hand, is one whose meaning does not apply to itself, such as 'French' or 'monosyllabic', for 'French' is not a French word, and 'monosyllabic' is not a word of a single syllable.

'Self-descriptive' and 'non-self-descriptive' are mutually exclusive adjectives, and so all words must be described by just one or the other. The paradox arises when we consider whether the word 'non-self-descriptive' is a self-descriptive or non-self-descriptive word. This is a semantic analogue to the set of all sets that do not contain themselves—it is the word that describes all words that do not describe themselves. If we contend that this word is self-descriptive then it is non-self-descriptive, and if it is non-self-descriptive, then it is self-descriptive. This is the same paradox Russell identifies for mathematics, yet we cannot solve it by saying that the word 'non-self-descriptive' is undefinable and meaningless, so why should we think that we should do so for Russell's?

Self-referential paradoxes have had a place in the Western discourse since long before the first academy. The oldest of these known paradoxes comes from a Cretan named Epimenides around 700 BCE, who proclaimed 'All Cretans are liars.' If this statement is true, then Epimenides, being a Cretan himself, is lying, and so not all Cretans are liars, and the statement is false. It is not a true paradox in this form, however, since the negation of the universal quantification of 'Cretan'—'all'—is the existential quantifier—'some'. If not *all* Cretans are liars, then there is at least one Cretan who is honest, and so we can conclude that the statement is false without contradiction: not all Cretans are liars, but Epimenides is. The paradox was clarified a couple of centuries later—around the time of Aristotle—by the philosopher Eubulides into the now famed *liar paradox*, which removes the quantification. It states, 'What I am saying now is a lie', or, as it is more commonly posed, 'This statement is false', which if true is false, and if false is true.

Despite its shortcomings, Russell and Whitehead's theory of types enjoyed considerable prestige as a foundation for mathematics in the early 20th century, for it did indeed avoid the paradox while allowing the derivation of the axioms of arithmetic. However, in comparison with Frege's theory, type theory was hard to justify as being grounded in pure logic. Russell's stratification of the universe of discourse and his vicious circle principle required additional axioms to maintain a satisfactory account of the concept of number, and it is not immediately clear that these are logical laws.[148]

Russell eventually came to have his own doubts, and soon the difficulties faced in altering the language of logic became more undesirable to mathematicians than abandoning a strictly logical foundation. Resolutions to the paradox which retained more of the original flavour of set theory grew increasingly popular, and most popular of all was that of the German mathematician Ernst Zermelo. Before we move on, it is worth briefly mentioning Zermelo's theory here as it would be his theory that solidified its place as the standard foundation for mathematics to this day.

Axiomatic Set Theory

Zermelo apparently had discovered the paradox of self-reference himself shortly before Russell, and he had begun work on his own way of circumventing it. His solution was not to alter the basic grammar of logic, as Russell had done, but to redefine what it means to be a set. In short, while Frege allowed any definable collection of objects to be a set, Zermelo only allows a definable collection of objects *within an existing set* to form a new set. Since the set comprehension axiom can now only pick out subsets of sets we know already, there is no way to collect all sets together, and so the universal set cannot exist. Of course, this by itself destroys set theory, for in order to define subsets we shall need at least one set that is not a subset itself, yet this we do not have. It is here that marks the end of Frege's logicist intentions of grounding mathematics in self-evident laws, because to reconstruct the perfectly rational and non-contradictory collections that are destroyed in the absence of Basic Law V, Zermelo must introduce a series of new rules.

Zermelo used a total of seven axioms in his initial formulation in order to *build* new sets. First, we need something to build them out of, and so Zermelo asserts the existence of at least one set: the empty set—the set which contains no members. From this, and by way of further axioms, we can build the set which contains the empty set, the set which contains the set which contains the empty set, the set of both these sets, and a whole range of larger sets still. Other axioms allow us to create sets of different types, such as power sets—sets which contain all the subsets of a set, along with the set itself—and thus, despite only being able to predicate of smaller sets with Zermelo's replacement for Basic Law V, we can *build* bigger ones by asserting additional rules. The result is a cumulative hierarchy of sets that is not all too dissimilar to Russell's hierarchy of types.

Since Zermelo rejected Frege's intentions of grounding mathematics in logic, there is technically no restriction on the axioms we can assert of set theory, provided their usefulness. The axioms are, therefore, postulates or assumptions rather than logical laws, and their purpose is simply to supply the Peano axioms without generating an inconsistency. Accordingly, there are many formulations of axiomatic set theory that we can use for the intended purpose, and indeed axioms have been added and removed from Zermelo's formulation over the years to arrive at the standard form of set theory used today: Zermelo-Fraenkel set theory with the additional axiom of choice, which was once dropped and later added again.

A notable issue of the theory concerns how we can understand the universe of discourse, for 'ZFC', as it is commonly called, does not allow this object to constitute a set. Of course, the universe must exist in any case because if a set exists, then it is a part of the collection of all sets. The question is, therefore: what is the collection of all sets? The standard response is that the universal collection is a *class*, but not a set. So now we ask, what is a class? Well, all sets are classes, but not all classes are sets, so the two are related, but they are not identical. This is as much as we can get from ZFC, however, as it does not formalise the universal class; all of its classes *are* sets, and there is no explicit difference between these kinds of collections.

John von Neumann did give a more formal definition of the kind of class which is *not* a set—a 'proper class'—as being a collection that is not a member of any other collection. However, a class which *is* a set—a 'small class'—*can* be a member of other collections, and so the distinction between a class and a set is still rather unclear. This is something we must simply accept in axiomatic set theory, for within the model the universal class cannot be governed by the same rules that govern sets, for else we would have Russell's paradox. Since the axioms concern sets and not classes, Russell's paradox can be interpreted as a consequence of conflating classes with sets. The distinction between them must be understood therefore in the context of avoiding Russell's paradox; we need a way to talk about the universal quantification of sets, but we cannot allow such an object to be a set itself. The functional difference between sets and classes therefore is simply that the rules of set theory do not apply to classes.

All this goes to show that modern set theory, while incredibly useful for mathematics, is not the set theory of Gottlob Frege, which was motivated by epistemological reasons above its usefulness. The motivation for axiomatic set theory, as it was for Russell's type theory, is to avoid confrontation with the paradoxical set, and this is not the same as resolving it. Whereas Frege wanted to build upwards from known principles to develop an ontologically informative account of the concept of number, Zermelo and his successors worked abductively to reduce our conjectures to the best possible model. This required preventing us from quantifying over properties, and so unlike Frege's logic, Zermelo's cannot express the relations of properties that are extensive in natural language. That is, in Frege's second-order logic, we can talk about any specific collection within a larger collection of objects, just as we can in English, while in Zermelo's first-order logic, we can talk only of the general collections.

Gödel's Incompleteness Theorems

In the year 1900, at the International Congress of Mathematicians in Paris, the German mathematician David Hilbert—one of the leading and most influential mathematicians of the era—presented a list of 23 unsolved problems of mathematics. In his second problem, Hilbert asked for a proof that the basic axioms of arithmetic will never lead into contradictory results, that the system be capable of "no further extension", and that no statement be regarded as true unless it can be derived from such axioms.[149] Hilbert would later clarify his program by defining a number of characteristics that any foundation of mathematics should have, and of central importance were those of *completeness* and *consistency*.

Completeness is satisfied when every true statement can be produced by the theory, and consistency is satisfied when every false statement cannot. In other words, all truths should be theorems, and all theorems should be truths. Completeness and consistency are two sides of a single coin, and their necessity for mathematics is implicit in our very attempts to uncover mathematical truth. The properties are also assumed for truth in the abstract, outside of formal systems, as they are implied by the laws of thought which ground all reason. The law of non-contradiction asserts that a statement and its negation cannot both be true—that truth is consistent—and the law of excluded middle asserts that a statement *or* its negation *must* be true—that truth is complete.

Truth and falsity are opposites, so the negation of every true statement should be false, and the negation of every false one should be true. Hilbert demanded that we demonstrate this for our mathematics, along with a method of determining whether any statement is a theorem or not—whether it is *decidable* by the theory. Similar considerations were also important at this time for epistemology and the philosophy of science, as the growing movement of logical positivism rejected the possibility of synthetic a priori knowledge and held that statements are only meaningful where they may be verified via experiment or logical proof.

In the late 1920s, Austrian mathematician Kurt Gödel was completing his education, and both Russell's type theory and Zermelo's set theory had been around for over two decades. It was generally recognised that both were successful in their aims of providing a foundation for mathematics while avoiding the paradoxes of self-reference. Hilbert's confidence that a simple set of axioms could combine in ever more complex ways that every true statement in mathematics could be produced, had been encouraged by Russell's theory, but it had not yet been demonstrated that "within every field of knowledge contradictions based on the underlying axiom-system are *absolutely impossible*".[150] Unbeknownst to Hilbert, it would be the young Kurt Gödel who would soon reveal that it's the demonstration itself which is demonstrably impossible, not the occurrence of contradictions.

In his study of the *Principia Mathematica,* Gödel gleaned an important insight—that despite explicitly banning the occurrence of self-reference, Russell's theory of types was ineluctably bound to it, for the implicit self-reference and self-justifying intentions of formal systems give rise to a disparity between what we can know about the system, and what it can 'know' about itself. By 1931, Gödel had devised a way of making this implicit self-reference explicit, revealing a vicious circle that was permissible by the vicious circle principle, and thus proving that self-reference cannot be simply avoided.[151]

Gödel's first step was to devise a method of expressing all of the theorems of type theory with a unique code number—a technique now referred to as 'Gödel coding'. He provides each of the symbols of type theory with an arbitrary positive integer; for example, the symbol '0' may be assigned the integer '2', and the symbol '=' may be assigned the integer '4'. These integers may be multiplied such that the formula '0=0' can be translated into the integer '16', being 2×4×2. There are of course many ways of generating the number 16 via multiplication, and Gödel needs each code number to be unique. Thus, he takes each symbol of the formula again and assigns them a second integer, this time not arbitrary, but a prime. The first symbol, '0', is given the first prime, 2; the second symbol, '=', is given the second prime, 3; and the third symbol, also '0', is given the third prime, 5. We now have two sets of integers—the arbitrary set: '2, 4, 2'; and the prime set: '2, 3, 5'. Gödel then raises the first set as exponents of the second, giving '2^2, 3^4, 5^2', and takes the product, which is a large but totally unique number, in this case '8100'.

The genius of this method lies in the fact that to decode any number we only need to find its prime factorisation, of which there will always be one. Some code numbers will be very large, as some formulas are large, but there will still be just a single way that any of these numbers may be decoded. This is the first step in Gödel's proof: allow each of the formulas of type theory to correspond to a unique number, which is their Gödel number. Any interaction of formulas in type theory are then paralleled by purely arithmetical calculations between these numbers, and any number which is reachable by these calculations will be the Gödel number of a theorem of type theory.

Once Gödel has established this method of encoding formulas, he is able to introduce a self-referential statement into type theory without violating any of the rules of the system. He uses as the template for this statement the oldest and most simple form of self-referential paradox—the liar paradox, 'This statement is false'. He translates this into a proposition of mathematical logic, equivalent to the sentence 'This statement is unprovable'.

We already know that in type theory no formula can contain itself, and a practical reason for this is that such a formula would be infinitely long. Consider in the natural language version that we insert the sentence into the referent within it; we would get '"'"This sentence is unprovable," is unprovable,' is unprovable," is unprovable…' ad infinitum. But now that all statements have a unique Gödel number, the self-referential Gödel sentence is equivalent to the proposition *G*: 'the formula whose Gödel number is *g* is not provable by the axioms of type theory'. We now have a statement valid in the language of type theory that presupposes itself, for the formula whose Gödel number is *g* is *G* itself.

We now want to decide whether this formula is valid. Can we prove *G*? Well, if we could then it would be true that *G* cannot be proved, since this is what *G* asserts, and we would have a contradiction. Furthermore, if we could prove *not-G*, or that *G* is false, then it would be true that *G* is provable, and so true—another contradiction. It therefore follows that *G* is not *syntactically* true or false in type theory—it is neither provably true nor provably false but rather *undecidable*.

The question remains, however: is *G actually* true or not? In other words, can we see that the statement must be valid by looking at it from outside of the system? Well, of course, we can.

The formula whose Gödel number is g is not provable in type theory, nor is its negation; this is precisely what we have just established. And since it is not provable, then Gödel's sentence must be true, and we have an example of a statement that we can see to be true, but which cannot be shown to be true within the system formally.

Now we have a troubling situation. Hilbert's program asserts that all true statements must be provable for an axiomatisation of arithmetic—that the system must be complete. Our only option is to now assert that G must be an *axiom* of the system and so is proved by default. Zermelo's set theory is grounded in extralogical postulates in any case, and so it is not altogether inadmissible that we should have to assert an additional axiom in type theory for purely logical reasons.

Let us call our original formulation of type theory 'T'. We now add G as an axiom to T and create a new system, 'T_2'. But here is the kicker; T and T_2 are different systems, and so the Gödel sentence of T_2 will have a different code number than the Gödel sentence of T. Accordingly, we shall have an entirely new Gödel sentence, 'G_2', stating: 'the formula whose Gödel number is g_2 is not provable by the axioms of T_2'.

No matter how we augment the system, there will always be a true but unprovable formula within it; and though we have used type theory as the medium through which to express Gödel's proof, the incompleteness theorem applies to any and all consistent formal systems that can generate a sufficient amount of mathematics. This is Gödel's *first* incompleteness theorem, and it asserts that if our formal system is consistent, then it is incomplete by necessity—there will be statements of its language which can neither be proved nor disproved within it.

Hilbert's program was therefore shown to be misguided on account of the inevitability of incompleteness, but what about consistency? Can we at least show within the system that the system will not give rise to contradictions? In this Gödel proves Hilbert mistaken again, on pain of contradiction. This is because we know from the first incompleteness theorem that *if a theory is consistent,* then there is a true but unprovable statement within it. Accordingly, any proof that the axioms are consistent would be a proof that the unprovable sentence *G* is provable, which would be a proof that the unprovable sentence *G* is unprovable, and a contradiction. Thus, the consistency of the system can only be proved if the system is in fact inconsistent.

Again, Gödel's proof only applies to consistent systems, and so providing that the system we are working in actually *is* consistent, and is grounded in classical logic, then the fact that we cannot prove its consistency means that it must be consistent, for inconsistent systems classically prove everything via explosion. Any statement purporting its own consistency therefore will be another example of a true but unprovable statement. This is Gödel's *second* incompleteness theorem—that any consistent and sufficiently powerful formal system cannot prove its own consistency—and it is a direct hit on the central aim of Hilbert's program.

Hilbert had reaffirmed his assurance that there are no unsolvable problems in mathematics and natural science in his retirement address, just a day after Gödel's first announcement of the first incompleteness theorem, which Hilbert did not attend. But Gödel's proof was not just a death blow to any hope for Hilbert's program, it also dismantled the main impetus behind the movement of logical positivism, with which Gödel himself was associated. Logical positivism relied on a principle of verification, which asserts that only statements justifiable via logical proof or empirical observation are cognitively meaningful. Gödel's proof presents a case where a true statement cannot be proved *within* the system it concerns, and so in regard to that system, verificationism entails that there are meaningless statements within the system that are meaningful outside of it.

Moreover, it's not simply the Gödel sentence, self-consistency, and other self-referential formulas that may be undecidable for a consistent formal system. There may be any number of statements of the language of arithmetic that cannot be given a truth value by, and are independent from, a given formal system. A mathematician may spend their entire life on a problem only for that problem to be impossible to solve. A notable case of just this occurred with the continuum hypothesis, which we mentioned briefly in Chapter Six.

The continuum hypothesis states that there is no set whose cardinality—number of members—is strictly between that of the integers and the real numbers. Georg Cantor, who founded set theory shortly before Frege, made it his life's mission to prove the hypothesis true, though others believed it false. It was in fact considered such an important issue that it was the first problem listed by Hilbert in his 23 unsolved problems of mathematics in 1900. After Gödel's proof, and after a lack of a proof that the hypothesis is true or false, the possibility became more apparent that this could be an example of a statement which is independent of Peano arithmetic. By 1963, it had in fact been shown that the hypothesis could neither be proved nor disproved in Zermelo's set theory and its derivatives.[152]

Chapter Fifteen

Gaps & Gluts

Intuitionism

Hilbert's desire to secure the completeness and consistency of arithmetic was connected to a position in the philosophy of mathematics called 'formalism'. In contrast to Frege's *logicism*, which held that mathematics derives from a real domain of logical objects, formalism contends that mathematics is merely a game of symbols. As such, meaning is found in our ability to manipulate the symbols in accordance with the rules of the game, and the truth of a mathematical statement is bound to its provability.

Hilbert's program was to find a set of axioms—a set of rules for the game—which gave rise to all truths and no falsities. Gödel's theorems contradicted the formalist view because for any set of rules identified, the consistency of that set of rules, along with its Gödel sentence, could never be demonstrated by it. That is, formalism cannot be successful because it is impossible for us to find a consistent set of rules that can prove all mathematical truths.

The lesson sometimes taken from Gödel's theorems is that they show mathematical truth to transcend mathematical proof, since statements we know to be true for a theory cannot be proven within it. However, the fact that we can know such statements to be true means they must be provable in some sense. It is rather that our ability to prove transcends what we can prove in any formal system. Mathematics is therefore non-axiomatisable in principle, for we will always need a metatheory to prove what cannot be proved within its object theory. This is no less troubling, however, for what it does show is that mathematical proof and mathematical truth do not always coincide—that if a statement is only provable in some metatheory, it is true but unprovable for the theory itself.

For some, this is too much of a bitter pill to swallow, for it calls into question our notion of proof, and the correspondence between syntax and semantics. Prior to the incompleteness theorems, we reasonably assumed that if something is unprovable then it cannot be considered true, and the idea that something can be *unprovably* true stands to contradict this view. It is in defence of our intuitions of the relation between proof and truth that the logician Ludwig Wittgenstein, a contemporary of Gödel, posed his highly controversial, yet historically misunderstood, critique on the consequences of Gödel's theorems in his *Remarks on the Foundations of Mathematics*.

Wittgenstein could not accept that there was a disparity between proof and truth in mathematics, for he believed that the meaning of a true sentence is delivered precisely by its proof. It is incoherent that we might have an unprovably true sentence because the truth of that sentence is provided by the system in which it exists.[153] Wittgenstein rejected the idea that a metatheory can be the source of meaning for the theory, for he saw no clear distinction between metamathematics and axiomatic arithmetic. A statement may therefore be provable and true, or not provable and not true, but not a mixture of them both.[154] To put this another way, the theory in which a statement is proved is a part of the sense in which that statement is true—it is true *in regard to that theory*—and thus Wittgenstein maintains that there cannot be statements that are true in regard to the same system in which they cannot be proved. Note also that while the Gödel sentence of a theory is true in some metatheoretic models, in others it may be false.

I mention Wittgenstein's views here with ambivalence, and as part of a discussion of the philosophical import of Gödel's theorems if we take for granted that truth is equatable with proof. We will find that the possible consequences of this view can be studied syntheoretically, in much the same way as we can discuss the possible theories of epistemology and ontology. What is for certain is that Gödel forces us to reconsider what we mean when we call something true, and Wittgenstein can be interpreted as foreshadowing the possibility that it is not our notion of truth that needs revision, but our approach to logic itself, and this leads us to the consideration of non-classical logics.

Wittgenstein's perspective is best understood by his constructivism, and his aversion towards the kind of mathematical Platonism that motivated Frege's logicism. For Wittgenstein, the symbols of mathematics do not reflect anything more real than our own conceptions, and so it is simply fallacious to talk about mathematical truth in the absence of them. There are, therefore, no truths in mathematics that are not known by mathematicians, and whenever new knowledge is gained, new truths are accordingly constructed. If this is indeed the case, then classical logic has been contradicted, for in the absence of knowledge of how a proposition is constructed, that proposition can be neither true nor false.

We find in this view some recourse in the philosophy of mathematics, for Frege's logicism had been proved inconsistent, Hilbert's formalism had been proved incomplete, and the question as to the nature of mathematics remained still wide open. It is widely recognised that Wittgenstein was influenced in this regard by the Dutch mathematician L.E.J. Brouwer, who asserted that mathematical objects are only real insofar as they have been constructed by human thought.[155] Without external truths to which our constructions correspond, a mathematical truth can only be held to exist when there is someone in the world who knows it.

As opposed to formalism, Brouwer's *intuitionism* survives Gödel's theorems because it does not presume that all statements should be decidable. Even if we accept the metatheoretical truth of the Gödel sentence, the intuitionist is safe for they are not held to the view that mathematics should be axiomatisable into a single system. So long as a statement is justified via some constructive means, the truth-conditions of that statement are satisfied. It is in those statements that have not been constructed, such as the continuum hypothesis, that intuitionism diverts from classical logic, for while the latter maintains the bivalence of undecided propositions—that they are true or false regardless of whether we can know if they are true or false—intuitionism denies this insofar as truth is conformant to what we know already. Things can *become* true or false, but they are not true or false by default.

The implication of taking an intuitionistic view on mathematics is that it forces us to alter the basic rules of logical consequence, which have long been held as forming the bedrock of the 'one true logic'. Specifically, in so far as undecidable statements are neither provably true nor provably false, the law of excluded middle must be denied. This has further consequences for our proof procedures as it prevents us from reasoning from false information. In classical logic, if we know that a statement is false, then we can infer that its negation is true, but this is not so in intuitionistic logic. The negation of a falsehood is not necessarily a truth, intuitionistically, and so a double negative is not the same as a positive.

Dialetheism

There is, however, another interpretation that can be made of Wittgenstein's remarks, and it follows when we take the identification of proof and truth not to mean that the unprovability of the Gödel sentence necessitates its untruth, but rather that its metatheoretically demonstrable truth necessitates its provability. Francesco Berto, in his commentary on Wittgenstein, proposes that Wittgenstein's intentions were to argue not that the metatheory must be eradicated in providing meaning to the theory, but that it must be considered a part of the theory itself. He therefore claims that Wittgenstein sought to frame Gödel's theorem as a genuinely paradoxical statement.[156]

Berto utilises an argument given by the British logician Graham Priest—a leading figure in one solution to the paradoxes of self-reference—whereby the reference to provability in the Gödel sentence means simply being able to establish the claim as true, rather than being able to derive it from certain axioms. Priest refers to this as a *naive* notion of proof. That is, if the metatheoretical provability of the Gödel sentence is taken as a condition for its provability simpliciter, then 'This statement is not provable' is false, since it is provable that it is not provable, which also makes it true. Priest explains:

> In this context the Gödel sentence becomes a recognisably paradoxical sentence. In informal terms, the paradox is this. Consider the sentence 'This sentence is not provably true.' Suppose the sentence is false. Then it is provably true, and hence true. By reductio it is true. Moreover, we have just proved this. Hence it is provably true. And since it is true, it is not provably true. Contradiction.[157]

That a theory gives rise to contradictions is classically evidence that somewhere we have made a mistake, and the mistake usually taken from this image is that we have incorrectly assumed that proof coincides with truth. Priest counters this judgement by proposing that our mistake is rather that we have assumed that contradictions cannot exist, and that Gödel's theorem should rightly be taken as showing *either* that truth exceeds proof *or* that contradictions exist. It appears that even Bertrand Russell, who sought to preserve consistency by simply sidestepping his paradox, may have thought through this possibility as well, for in a letter of 1963, he writes:

> It is fifty years since I worked seriously at mathematical logic and almost the only work that I have read since that is Gödel's. I realised, of course, that Gödel's work is of fundamental importance, but I was puzzled by it. It made me glad that I was no longer working at mathematical logic. If a given set of axioms leads to a contradiction, it is clear that at least one of the axioms must be false. Does this apply to school-boys' arithmetic, and, if so, can we believe anything that we are taught in youth?...You note that we were indifferent to attempts to prove that our axioms could not lead to contradictions. In this Gödel showed that we had been mistaken.[158]

Whether Russell was humouring Gödel's proof as an inconsistency theorem in this remark, or whether he was conflating truth and proof unwittingly, is unclear. Nevertheless, as Russell's words express, following Gödel's theorems, philosophers began to take less of an interest in the foundations of mathematics, in lieu of the loss of epistemological security begot by the acceptance of incompleteness. Speaking generally, mathematicians are naturally inclined to favour consistency over completeness, for, classically, any contradiction trivialises our theories, rendering them useless.

Nevertheless, on the realist and naive view of truth, regardless of whether we have a proof, any statement must be true or false, and if we should know anything at all, we should be able to know it all. "There is no ignorabimus," to quote David Hilbert, who felt this sentiment so strongly that he had it carved into his gravestone. Wittgenstein too, as Berto sees it, insisted that all mathematical propositions must be decidable.

It is therefore necessary that we consider the opposing view: that completeness is more desirable than consistency, and that just as the intuitionists have altered logic to deal with propositions being neither true nor false, we should consider logics that can deal with propositions being both. Of course, this is a serious matter for it requires the violation of Aristotle's 'most certain of all principles', the law of non-contradiction. Priest opens his influential 1979 paper, 'The Logic of Paradox', with a quote from Wittgenstein which explicitly addresses the possibility of true contradictions: "Indeed, even at this stage, I predict a time when there will be mathematical investigations of calculi containing contradictions, and people will actually be proud of having emancipated themselves even from consistency."[159]

Despite the obvious and concrete irrationality of accepting contradictions into our logic, there are reasons why doing so may be favourable over accepting the converse—the absence of truth-valuation altogether. First, there is the epistemic advantage that an inconsistent theory can fully model its own semantics, and then there is the pragmatic advantage that it does not prevent us from inferring a truth from the negation of a falsehood.[160] On the other hand, it does prevent us from inferring a falsehood from the negation of a truth, but this is less useful for the derivation of new theorems. The most interesting advantage of having an inconsistent logic, however, is it that it would allow us to re-establish Frege's logicism and the unrestricted comprehension of sets defined by his Basic Law V.

We are reminded that Frege's foundation of arithmetic was deemed to fail because it entailed the existence of the Russell set—the set of all those sets which do not contain themselves—violating the law of non-contradiction. Russell's and Zermelo's logics were designed to circumvent this set, and thus the paradox, but in so doing they fractured themselves from the logic of natural speech. If we were to admit of contradictions, however, then we can assert that the Russell set both contains itself and not, reviving logicism along the way. It would also allow us to resolve other classically undecidable propositions, such as the continuum hypothesis or the Goldbach conjecture, and furthermore to form a complete axiomatisation of arithmetic based on self-evident principles, where every truth is provable, along with some of their negations.[161]

Indeed, Frege believed that Basic Law V was an analytic law of logic, and if this position is to be retained in spite of Russell's paradox, we will have arrived at the conclusion that contradictions exist in any case. Priest refers to such contradictions as 'dialetheia', and the belief in them as 'dialetheism'. This is a fitting name for it provides the antithesis to that position we have labelled previously as 'monoletheism'. Dialetheia are taken as a kind of double-sided object, existing at the intersection of truth and falsity and staring in both directions. This too was foreshadowed by Wittgenstein when he said:

> Why should Russell's contradiction not be conceived of as something supra-propositional, something that towers above the propositions and looks in both directions like a Janus head? The proposition that contradicts itself would stand like a monument (with a Janus head) over the propositions of logic.[162]

The main reason that the introduction of contradictions is classically incoherent is that they entail the truth of any proposition, in accordance with the principle of explosion—*ex falso quodlibet*. If *A* is both true and false, then for any statement *B*, *A* or *B* is true, and since *A* is false, we can infer that *B* is true. Yet, *B* can be any statement whatsoever, and so classically a contradiction results in trivialism, which would make dialetheism incoherent.

The dialetheist response to this is that if the fact that *A* is false does not exclude the possibility that *A* is also true, there is no reason that we should infer *B* from *A* or *B*. To put this another way, *A* is not simply true, and it is not simply false; rather, *A* is true *and* false, and this acts as a third truth value. In this way, we can argue that the principle of explosion is only self-evident if we are assured that our domain of discourse is monoletheic. Once more, this too was foreshadowed by Wittgenstein: "One may say, 'From a contradiction everything would follow.' The reply to that is: Well then, don't draw any conclusions from a contradiction; make that a rule."[163]

Of course, if we deny the law of non-contradiction completely then we allow that *any* proposition could be both true and false, implying a kind of triviality in potentia, which we do not want. It is a bivalent fact of the matter as to whether, for example, all widows are women, and so we do not want to allow 'true and false' to be a potential truth value for this proposition. In fact, dialetheism has no need to actually deny the law of non-contradiction, for it can simply contradict it and assert that the law is a dialetheia itself. The crucial alteration of classical logic is rather the denial of the principle of explosion, which contains the law of non-contradiction in the form of a consequence and is therefore stronger than it. What the principle of explosion tells us is that if we are to accept anything as false only, then we must also accept there are no contradictions.

Paracompleteness & Paraconsistency

We have now seen that there are two obvious approaches to saving our common-sense notions of proof and truth in mathematics. We can assert that truth conforms to constructive proof and that some statements are neither true nor false, or we can assert that proof conforms to truth and that some statements are *both* true and false. Each of these options require the alteration of classical logic, and so classical logic no longer holds the incontrovertible position it once had—it can be rejected. A logic which denies the law of excluded middle is called a 'paracomplete logic', being incompleteness-tolerant in that all true statements are provable, yet not all statements or their negations are true. A logic which denies the principle of explosion, on the other hand, is called a 'paraconsistent logic', being inconsistency-tolerant in that some true statements are also false, but the contradiction does not entail trivialism.

Inconsistent logics are not susceptible to Gödel's incompleteness theorems, which take consistency as a premise, and so there is no reason why they cannot be complete—capable of proving all well-formed statements or their negations. Incomplete logics, conversely, cannot decide all propositions, but they need not contain any contradictions. We can therefore see that logic is bound by a principle of complementarity between the conflicting notions of completeness and consistency. We cannot achieve both within a single theory, yet both are required for a full account of logical consequence. The laws of thought, too, can be seen as complementary in the context of non-classical logic. Paraconsistent logics abandon consistency while retaining completeness, violating the law of non-contradiction while observing the law of excluded middle. Paracomplete logics abandon completeness while retaining consistency, neglecting the law of excluded middle while upholding non-contradiction.

The duality between these two types of logic can be defined formally, and research on this topic has been led by a number of influential logicians from South America. Walter Carnielli explains: "From the point of view of classical logic, the validity of excluded middle in paracomplete logics and the invalidity of explosion in paraconsistent logics are like 'mirror images' of each other." Here, Carnielli is taking the law of excluded middle in the conditional form of logical consequence—'if *B*, then *A* or not-*A*'—where the premise is any valid formula, and the consequence is a tautology defined by the law of excluded middle. In this form, it is sometimes referred to as the principle of *implosion*, just as the consequential form of the law of non-contradiction is the principle of explosion. We can clearly see the duality between the two principles, for when we switch either side of the implication and negate the logical connectives, they are transformed into each other. $B \vdash A \vee \neg A$ becomes $A \wedge \neg A \vdash B$ and vice versa.[164]

The duality also applies to epistemological paradigms in regard to the philosophy of science. Constructive paracomplete logics such as those based on Brouwer's intuitionism provide a possible line of defence for positivism in light of Gödel's theorems. Michael Dummett for one has asserted that verificationism leads to intuitionistic logic, and intuitionism is inherently verificationist for it only accepts formulas as valid where they are built from other truths.[165] On the other hand, paraconsistent logics have been presented as the proper foundation for the complementary *negativist* methodological doctrine of falsificationism. David Miller identifies the strict dual of standard intuitionistic logic to demonstrate this duality in the philosophy of science.[166]

There are many varieties of paraconsistent and paracomplete logics, some closer to classical logic than others, but each logic within a hierarchy of paracomplete logics will have a paraconsistent dual, and vice versa. Carnielli and Brunner have clarified the paracomplete–paraconsistent relation to verification and falsification respectively by developing such hierarchies and defining the method for their dualisation. Just as "the intuitionistic philosophic program is committed to constructing truthhood" the dual anti-constructive paraconsistent logics identified "can be seen as committed to eliminating falsehood".[167] In this respect, the fact that paracomplete logics can have contradictory propositions simultaneously false can be appropriated for a logic whose propositions may be considered false until proven true. Similarly, the fact that paraconsistent logics can have contradictory propositions simultaneously true can be appropriated for a logic whose propositions may be considered true until proven false. The complementary bias of truth-valuation defaults helps us to understand the duality between logics, since classical logic has no such bias and so its dual is identical to itself.

Expanding on the idea of truth-valuation defaults, consider that we have a set of propositions that are undecidable in a given system. This may be the Gödel sentence of the system and its negation, the continuum hypothesis and its negation, or any other independent propositions. In a paracomplete theory with a falsity-default, these statements may be called false insofar as they cannot be verified, while in a paraconsistent theory with a truth-default, they may be called true insofar as they cannot be falsified.

Classically, we can define truth and falsity in both the positive and the negative, which is to say, a true statement can be one with a false negation, or it can be non-false, while a false statement can be one with a true negation, or it can be non-true. Non-classically, the positive and negative descriptions are no longer equivalent, since the laws of excluded middle and non-contradiction, and the principles of implosion and explosion, are required to define classical negation. A truth in paraconsistent logics need not have a false negation, and a falsity in paracomplete logics need not have a true one.

Theoremhood

We can characterise the duality between a consistent paracomplete system and a complete paraconsistent system by considering the relationship between the theorems and nontheorems of those systems, which is to say, the statements provable in the system and the statements not provable. These two sets should complement each other in the sense that the information encoded by the set of theorems should be a perfect reflection of the information encoded by the set of nontheorems. This is because, classically, if we have a proved statement—a theorem—and we take its negation, we have a nontheorem, and if we take a disproved statement—a negated theorem—and we take its negation, then we have a theorem.

To visualise this, consider the relation between a relief and intaglio sculpture, which are two different ways of representing the same information in a physical medium, such as a stone or wood carving. In relief sculpture, the background of an image is carved away from the material, and the figure is projected outwards, while in intaglio sculpture, the figure is carved out of the surface such that it appears hollow beneath the background. Consider that we have two blocks of wood, each encoding precisely the same figure and shape. On the first, the background is carved out in the relief style, and on the second, the figure is carved out in intaglio. If we have done the job perfectly, these two pieces of material will now be able to fit snugly inside of one another, without leaving any space between them. The shape of each carving encodes precisely the same information, the only difference being that it is inverted between them, one positive and one negative.[168]

If we take for granted that we have found proofs for all provable statements, the theorems of a system can be analogised to the relief carving, projecting out of the system, while the nontheorems can be analogised to the intaglio carving, recessing inwards. However, as we have discussed, this analogy does not work for sufficiently powerful systems of arithmetic. If that system is consistent, there will be statements of its language which are neither provable nor disprovable from its axioms, and there will be nontheorems which are not the negations of theorems. In other words, the set of disprovable statements is smaller than the set of unprovable statements, considering of course only well-formed strings that are interpretable as mathematical formulas.

To reiterate the point, neither the Gödel sentence nor its negation is decidable in a given consistent system, and so both are nontheorems of that system, and neither are negations of theorems. This is true in classical logic but also in paracomplete logics, and from this we can conclude that the set of theorems of a paracomplete system contains *less* information than its set of nontheorems. This is analogous to the relief carving being ever so slightly smaller than its intaglio equivalent, such that there is a gap in between them. The notion of gaps is important to paracomplete logics because when we take the notion of proof as equivalent with truth, those statements undecidable by the system can be said to fall into a 'gap' between truth and falsity. When we try to bifurcate true statements from false ones, this gap falls onto the side of the nontheorems.

The reverse situation occurs when we take the intuitive demonstration of the Gödel sentence to be a condition of its provability, such that Gödel's proof becomes a paradox as described previously. Here, the Gödel sentence is provable if it is unprovable and unprovable if it is provable, and this paradox can be dealt with if we accept the sentence as both true and false, asserting a paraconsistent logic. That is, 'this statement is unprovable' is a theorem, and its negation is also a theorem, such that neither are negations of nontheorems. In such a paraconsistent logic, the set of theorems contains *more* information than its set of nontheorems, and the analogy here is a relief carving being ever so slightly *larger* than its intaglio counterpart. Here, there is *too much* material in the former to sit snuggly inside of the latter, and we have a 'glut' between truth and falsity, whereby the intersection between them is not empty. When we try to bifurcate true from false statements, this glut falls onto the side of the theorems.

The domain of proof is like a big jigsaw puzzle, which we have filled save for the final piece. We have one more piece available, but this piece is a conjoined twin—it is twice the size of the final space. We have a choice, therefore, between leaving out this final piece, or overlaying the double-piece onto the single space. The former option results in a gap in the image, whereby we do not have enough information present for a complete description of the world, but the pieces that we do have fit tidily together in our image—this is the monoletheic approach of consistency over completeness. The latter option results in a glut in the image, whereby we have too much information present to maintain a tidy image, but we do have enough information to describe everything within the universe of discourse—this is the dialetheic approach of completeness over consistency.

Now, let us informally consider a dual pair of systems, one based on a verificationist paracomplete logic, '*M*', and one based on a falsificationist paraconsistent logic, '*D*'. Considering only well-formed formulas, each logic has a set of theorems and a set of nontheorems which for the most part look the same, diverging only when they meet a classically undecidable proposition. These propositions cannot be separated from their negations and so they must either be considered valid together, or invalid together. *D*, being falsificationist, accepts these double-sided propositions, categorising them into its set of theorems, while *M*, being verificationist, rejects them and places them into its set of nontheorems.

As we have seen, in both of these logics, the sets of theorems and nontheorems are not perfect opposites, and so do not contain the same information. Both logics encounter pairs of propositions, one the negation of the other, and how they deal with these propositions determines whether the logic will be inconsistent or incomplete respectively. Inconsistency arises when the contradiction is positive, while incompleteness arises from a *negative* contradiction, which is a violation of the law of excluded middle. Only when we consider *M* and *D* together, as parts of a greater system, do our theorems and nontheorems return into balance. That is, the set of theorems of *M* is just the negation of the set of nontheorems of *D*, and the set of theorems of *D* is just the negation of the set of nontheorems of *M*.

M and *D* are inversions of each other, and this duality

requires that completeness and consistency be complementary characteristics of our systems in general. To return to our analogy of relief and intaglio sculpture, the fact that two seemingly opposite shapes are too big or too small to fit inside each other can be resolved if we acknowledge that we are missing half of the arrangement. Perhaps, the problem faced in classical logic, with our inability to separate certain self-referential statements from their negations, is that such statements are not separable in principle. Perhaps, paradoxes express one idea appearing as two in opposition, and our common sense and language are just incapable of expressing this interdependence of opposites. We can accept them both, or we can reject them. There is no clear answer as to which of the two it should be, and so paracompleteness and paraconsistency are antitheses of non-classical logic just like those in other areas of philosophy. If we can show that this dialectic is definable by the dialectical matrix, then we can show that the rationale of discrimination in philosophy is governed by that which it seeks to govern—we can show that discrimination in principle is fallacious.

Chapter Sixteen

On Logical Consequence

Theoremhood Continued

In the previous chapter, we discussed the inevitable asymmetry that occurs between the theorems and nontheorems of formal systems. Our intuition that what we can know to be true should reflect what we cannot is mistaken, for there will always be more contained in one or the other. Self-reference explains to us why this must be the case, for certain self-referential propositions are inseparable from their negations, and they require non-classical logic to naively account for their valuation. Paracompleteness allows a statement and its negation to not be true, and so a paracomplete theory may have nontheorems that are not the negations of theorems. Paraconsistency allows a statement and its negation to not be false, and so a paraconsistent theory may have theorems that are not the negations of nontheorems. These two approaches are duals of each other, so that the theorems of the former are negations of the nontheorems of the latter, just as the theorems of the latter are negations of the nontheorems of the former.

We are now in a position to consider the structure of non-classical perspectives on formal proof syntheoretically, in much the same way as we distinguished the different perspectives on knowledge, being, and value in Part II. There we identified a quadrant diagram related to each of these notions, referred to as the dialectical matrix, which was defined by a complementary pair of properties along with their negations. That is, the positive property 'determinateness' and the negative property 'immutability'. When considering the notion of proof, there are also four syntactical values of a system: 'provable', being positive; 'disprovable', being negative; and their negations, 'non-provable' and 'non-disprovable'.

For now, we will assume that we are working within a system of classical logic, so we shall not conflate the notion of proof with the notion of truth. In such systems, not all propositions are provable or disprovable, but all are provable or non-provable, just as they are disprovable or non-disprovable, and so all propositions exist within two dimensions of provability and theoremhood. Another way to put this is that not all formulas will be theorems or antitheorems—negations of theorems—but all will be theorems or nontheorems, just as they will be antitheorems or non-antitheorems. These four properties can be expressed on the dialectical matrix as follows:

	Nonprovable	Provable
Non-disprovable	$[\neg T + G] \cap [T + G]$ G = neither	$[T] \cap [T + G]$ T = true only
Disprovable	$[\neg T + G] \cap [\neg T]$ $\neg T$ = false only	$[T] \cap [\neg T]$

Accordingly, the interaction of these dimensions forms four composite proof values: 'provable and non-disprovable', 'disprovable and non-provable', 'non-provable and non-disprovable', and 'provable and disprovable'. Again, taking classical logic as a baseline, we can categorise all of the propositions of a given theory into this diagram. If that theory is an axiomatisation of arithmetic, such as a set theory, then the self-referential Gödel sentence and its negation will be examples of statements that are neither provable nor disprovable within it, and there will be no statements that are both provable and disprovable. Let us symbolise all provable statements as 'T', all disprovable statements as '$\neg T$', and all undecidable statements as 'G'. Note that everything disprovable will also be non-provable, and everything provable will be non-disprovable. As such, our four categories of propositions are: provable statements, T; non-provable statements, $\neg T$+G; disprovable statements, $\neg T$; and non-disprovable statements, T+G.

Now, suppose we take proof to be equivalent to truth, and use the conjunctions of these categories to form the foundation of four perspectives on non-classical logic. The upper-left class designates G as 'non-true and non-false'; the upper-right class designates T as 'true and non-false'; the lower-left class designates $\neg T$ as 'non-true and false'; and the lower-right class designates nothing as 'true and false'. That is, the propositions contained within the conjunction of two respective categories are designated a different truth value within each of the four quadrants. We can then use this framework to derive four kinds of non-classical logic.

We can begin to see already that the lower-left quadrant is going to be related to a falsificationist paraconsistent logic, given that its designated value is 'non-true and false', while the upper-right quadrant is going to be related to a verificationist paracomplete logic, given that its designated value is 'true and non-false'. This is made clearer by taking those propositions not designated with a truth value within a class to be defined by the negation of the truth value that *is* designated. The upper-right class designates T as 'true and non-false', and so $\neg T$+G will be 'non-true or false', which is consistent with verificationism and paracomplete logic. Similarly, the lower-left class designates $\neg T$ as 'non-true and false', so T+G defaults as 'true or non-false', which is consistent with falsificationism and paraconsistent logic.

	Non-true	True
Non-false	Non-true & Non-false G True or False $T + \neg T$	True & Non-false T Non-true or False $\neg T + G$
False	Non-true & False $\neg T$ True or Non-false $T + G$	Non-true & Non-false G True or False $T + \neg T$

We are particularly interested in the upper-right and lower-left classes, partly because these reflect paracompleteness and paraconsistency respectively, but also because they are correlated with the syntheoretic belief systems that we labelled previously as 'objectivism' and 'subjectivism'. We can see that the propositions of these classes are like mirror images of each other. The *false and non-true* propositions ($\neg T$) of the lower-left class are negations of the *true and non-false* propositions (T) of the upper-right, just as the *non-false or true* propositions (T+G) are negations of those that are *non-true or false* ($\neg T$+G). Note that the negation of G is just G since these propositions are already coupled with their negations. The positioning of these classes also makes sense syntheoretically, since the subjectivist ideology is related to positions such as idealism, logicism, Platonism, and falsificationism, while the objectivist ideology is related to positions such as materialism, intuitionism, nominalism, and verificationism. We can therefore now associate a preference for completeness with subjectivism, and a preference for consistency with objectivism.

The Structure of Logic

We have now identified four kinds of logical systems, but these are not *formal* systems since we are not starting from a set of axioms and deriving valid formulas. Rather, we are starting from sets of designated formulas and assessing what kind of logics they must be formulas of. We can see that there is a duality in the designation of truth values, and so we can infer the logical principles that would give rise to such designations when we take proof as equivalent to truth. This is possible because there are *four* key principles that can be denied of classical logic to give rise to various kinds of non-classical logic, and these principles can be characterised by the dialectical matrix.

The distinction between paracompleteness and paraconsistency is typically characterised by the fact that the former rejects the law of excluded middle, while the latter rejects the principle of explosion. However, we also know that inconsistent logics violate the law of non-contradiction, and so a logic which denies the principle of explosion *and* the law of non-contradiction has been referred to as a 'strong' or 'genuine paraconsistent logic'.[169] Genuine paracomplete logic has also been defined as a logic which denies the duals of these principles: the principle of implosion and the law of excluded middle. Here, the principle of implosion is presented as the contrapositive of $B \vdash A \vee \neg A$, which is $\neg(A \vee \neg A) \vdash \neg B$.[170] The former states that a proposition or its negation follows from any other proposition, and the latter states that the negation of a proposition follows from a 'gap'. The principle is sometimes referred to as 'weak explosion' and is invalid in most paracomplete logics since they allow for $\neg(A \vee \neg A)$ and so $A \vee \neg A$ is not a tautology. Our four principles are thus:

$\vdash\neg(A\wedge\neg A)$	**(LNC)**	**not *'A and not-A'***
$\vdash A\vee\neg A$	**(LEM)**	***A* or *not-A***
$A\wedge\neg A\vdash B$	**(EXP)**	**if *A* and *not-A*, then *B***
$\neg(A\vee\neg A)\vdash\neg B$	**(IMP)**	**if not *'A or not-A'*, then *not-B***

The first pair express rules that should hold for all propositions, and the second defines the consequences of breaking those rules. The four principles can now be used to define four classes of non-classical logic by using them as categories of the dialectical matrix. This process is relatively straightforward, for we already know which quadrants will result in paracompleteness and paraconsistency respectively, but there are also syntheoretic reasons for which positions on the matrix the principles should take. One way to see this is by considering their correlations with the categories of justification in epistemology.

Intuitionism and formalism depend on the premise that the laws of logic are intuitions, postulates, and therefore *synthetic*, and so the characterising law of thought for paracomplete systems—the law of non-contradiction—is syntheoretic with the epistemological category 'synthetic'. The law of excluded middle is then related to the category 'analytic', being necessary for any logicism, but also because $A\vee\neg A$ is a classical tautology. The principle of explosion must then be syntheoretic with the category 'a posteriori', which makes sense since it is an a posteriori observation of the empirical world that not just everything is true, and so there must not be contradictions. Finally, the principle of implosion must be related to the category 'a priori', which may be explained by the fact that a priori knowledge would be impossible if it were not the case that a proposition or its negation is indeed true. We can now represent these four principles upon the dialectical matrix as so:

	$\neg(A \vee \neg A) \vdash \neg B$	$A \wedge \neg A \vdash B$
$\vdash \neg(A \wedge \neg A)$	Non-true & Non-false Implosion Non-contradiction Paraconsistent Paracomplete	True & Non-false Explosion Non-contradiction Consistent Paracomplete
$\vdash A \vee \neg A$	Non-true & False Implosion Excluded Middle Paraconsistent Complete	True & False Explosion Excluded Middle Paradoxical Trivial

With these rules established, we are now in a position to inquire into the remaining two classes, which are syntheoretic with *abjectivism* at the top-left, and *superjectivism* at the bottom-right. We will find that these are unlikely to be logics we would consider developing for pragmatic purposes, in the form of formal systems, for they are quite unusual. Unlike the two biased classes, which express congruent conjunctions of truth values—'true' and 'non-false' for the paracomplete class; 'false' and 'non-true' for the paraconsistent class—the remaining classes combine opposing valuations—'non-true' and 'non-false' for the abjectivist class; 'true' and 'false' for the superjectivist class.

The abjectivist class does not feature any determinate valuations, for it *denies* what is true or false but does not *assert* anything. We cannot infer the negations of any denials since the class does not have the law of excluded middle as a rule. It also does not have the principle of explosion as a rule, which means that the class is simultaneously paraconsistent and paracomplete. Nevertheless, the principle of *implosion* is valid in the logic, and it also contains some statements—*G*—that are neither true nor false. Since implosion states that any negated proposition can be inferred from a truth value gap, the logic is weakly explosive, or rather, it is *implosive*, which is to say, we can infer any $\neg B$. To fully understand what this entails, more of the logic would need to be developed. Classically, it would mean that any negated proposition is true, but this is not classical logic. In this logic, 'not *B*' means that is not true, not that it is false, and so the result of this logic is that it denies everything that can be asserted—a complete *implosion* of truth valuation.

The superjectivist class, at the bottom right, appears to be even more strange. Here, the law of excluded middle and the principle of explosion are accepted as axioms, and so the logic is neither paracomplete nor paraconsistent, but the law of non-contradiction and the principle of implosion are denied. The logic tells us what is true and what is false, yet not all statements are included in these sets. As such, some statements, namely *G*, are neither true nor false. However, the logic validates the law of excluded middle, which states precisely that all statements are true or false, and so we have a contradiction. Since we also observe the principle of explosion, we can then infer any arbitrary statement, so that every statement is true, making this a logic of pure trivialism.

However, the matter is not quite so simple, for the logic is explosive just in case there is a statement such that neither it nor its negation is valid, yet if the logic is explosive then all statements are valid. It seems therefore that there is a self-referential paradox arising within the structure of the system itself, for the violation of the law of excluded middle acts as the trigger of the principle of explosion, which is immediately patched over once it is triggered, making it untriggered. In other words, the logic explodes if it does not explode, and does not explode if it explodes.

It is clear that there is something not quite right about this logic, for the law of excluded middle is a part of the syntax of the system, while truth value gaps are a part of its semantics. It seems then that the system must be taking two different logics and presenting them as one, and indeed there are two different states that the logic appears to assert in superposition. In its non-exploded state, the logic is consistent and incomplete since it contains no contradictions, but not all statements or their negations are true. In its exploded state, it is inconsistent and trivially complete, since all statements are true. Due to the self-reference of the system, each state is the cause of the other and so depends on it, yet the system is never able to settle on either one. The only conclusion is that the system exists in both states just as the dialetheist would assert the Russell set to contain itself and not.

That said, there is a simpler interpretation to be made of the superjectivist class, though it also requires defining more of our logical apparatus. If we allow that '*G* is neither true nor false' means 'neither *G* nor not-*G*', then by De Morgan equivalence we have 'not-*G* and not not-*G*'. If we also allow double negation elimination, which is generally valid in logics with the law of excluded middle, then we have 'not-*G* and *G*'—a glut. However, since truth and falsity are independent and belong to different dimensions of the dialectical matrix, and since the logic is non-classical, it's not clear that this equivalence can be drawn. If it can be, then the logic is simply explosive and trivial, which would provide a direct dual to the abjectivist class regardless, which is implosive and sceptical.

The fact that this logic is implied by the dialectical matrix also plays into the self-referential and paradoxical nature of the superjectivist ideology, with which it is syntheoretic. In Part II, we derived several self-contradictory notions, which arise from the conflation of complementary properties that cannot be combined into a single theory. In epistemology, this was the conflation of 'a posteriori' and 'analytic'; in ontology, it was the conflation of 'concrete' and 'universal'; in axiology, 'subjective' and 'absolute'; and in political theory, it was 'libertarian' and 'collectivist'. We can now understand the semantic notions 'truth' and 'falsity', the syntactical principles of explosion and excluded middle, and even consistency and completeness, in terms of this complementary relation.

In our prior inquiries, we understood these complementary properties to be mutually exclusive in the sense that, for example, all analytic knowledge is a priori, and all a posteriori knowledge is synthetic, but the converse is not true since we can have knowledge that is synthetic a priori. We now see this reflected in the dialectical matrix for logic, for, in a classical setting, everything disprovable is non-provable, everything provable is non-disprovable, some things are non-provable and non-disprovable, but nothing is both provable and disprovable.

Tarski's Undefinability Theorem

We have seen that Gödel's theorems, which apply to systems of logic from which a certain amount of arithmetic can be produced, necessitate some formulas which cannot be determined as true or false within the system, but which can be seen to be true outside of it. If we wish to retain a classical account of logical consequence for mathematics, we must accept that truth transcends proof, and if we do not, we must alter our logic to allow the existence of gaps or gluts respectively. We might be inclined to say that this is simply a peculiar characteristic of the nature of natural numbers, having nothing to say about truth in regard to philosophical or scientific inquiry at large. We shall now see that this is not the case, for the transcendence of truth, or the actuality of non-classical logic, is essential to any formal inquiry into our world, no matter the domain that we are working in.

While classical logic has been forced to accept that there is some kind of schism between our ability to prove something formally, and the fact of that something being true, we still expect that what we can prove formally can be reliably *labelled* as true; it's just that the set of true sentences exceeds the set of provably true sentences. However, Gödel's work shows us that this assumption is unfounded, and the matter was explained formally by Alfred Tarski in 1933.

We know the concept 'theoremhood' is expressible in formal systems like set theory, for a sentence satisfies this concept just in case it can be derived from the set of axioms. We also know that the concepts 'non-theoremhood', 'antitheoremhood', and 'non-antitheoremhood' are expressible in such systems, but Tarski was interested in the concept of 'truth', and whether *that* can be expressed formally.

The consensus throughout the history of philosophy is that the notion of truth is intuitive, not relying on extraneous explanation, and that utterances are true when what they say are so. Our proclamations are justified just in case they correspond to the world; this idea goes all the way back to Aristotle and has remained largely unchallenged. As Tarski phrased it in his 1933 paper, "A true sentence is one which says that the state of affairs is so and so, and the state of affairs is indeed so and so."[171] Tarski formalises the rule into what is now commonly called the 'T-schema', or truth schema:

> *X* is true, if, and only if, *X*.
> T '*X*' iff *X*

If we want to claim that it is true that grass is green, the sentence 'Grass is green,' is true if and only if it is the case that grass is green, and there is no difference in stating the assertion with or without a reference to its truth. For us to be able to designate this or any sentence as true, all of the true sentences of our language must be implied by the T-schema, for the T-schema is simply the inclusion of the concept 'truth' in any language. However, an issue arises when we apply the T-schema to the liar paradox. 'This sentence is false,' is true if and only if that sentence is false, and since the predicate for the liar sentence, 'being false', asserts, '"This sentence is false," is false,' the T-schema for the liar paradox states: '"This sentence is false," is true, if and only if, "This sentence is false," is false,' which is clearly incoherent.

Now, Tarski set up his theorem in the language of arithmetic using Gödel coding, but it is not just mathematics in which we make proclamations of truth. Tarski highlighted that any language that is able to express the concept of truth will be able to produce a formal version of the liar paradox, and the logical consequences for that language will clearly be devastating. If that language is based on classical logic, then we can derive a contradiction from the paradox, and that contradiction will explode under the principle of explosion, proving all statements as true, and trivialising the language. That is, if we were to create a formalisation of English based on classical logic, it would be the case that all English sentences are true. Tarski's conclusion, and the result of his 'undefinability theorem' of truth, is that no classical formal language can express the concept of truth—contain a truth-predicate—for if it did, we could produce the liar paradox.

Tarski notes that we can have a metalanguage that defines truth for its object language, but this metalanguage would have to rely on a further metalanguage to define its own truth-predicate. However, this would produce a similar situation as to the one we would have in adding the Gödel sentence to the axioms of a classical set or type theory—the hierarchy of metatheories will always be incomplete.[172] The solution requires that we alter our definition of truth away from the T-schema, requiring that truth transcends any correspondence to an actual state of affairs. Yet, since natural languages, such as English, *do* contain the concept of truth and do not rely on a metalanguage, altering the definition of 'truth' would not work for any formalisation of natural language.

The derivation of the liar paradox in *any* sufficiently powerful formal language is therefore a serious issue. We cannot merely dismiss it like we might be inclined to do with the standard paradox in English. As American logician Jc Beall states (with Glanzberg and Ripley):

> Such things might seem to be "mere puzzles," in the sense of games one might play, but they are far more. They show some very basic principles governing truth, indeed, principles so basic and obvious they might look like logic, might turn out to be *false*...Given the seeming obviousness of basic principles of truth, the fact that they lead to inconsistency if we make some background assumptions about logic creates a kind of dilemma. Either some obvious principles about truth, or some equally obvious principles about logic itself, must go.[173]

As Beall explains, the only way out of the paradox without having to rely on a metalanguage requires the alteration of classical logic, and the two options we have at our disposal are paracompleteness and paraconsistency respectively. That is, we must introduce a status value other than 'true' or 'false' and ascribe it to paradoxical sentences. Three-valued logics containing a truth predicate and satisfying the T-schema for all sentences of their language are called 'Kripke-Kleene models', and the manner in which this third truth value is interpreted determines the syntax of the resultant logic.

The common theme of a third truth value is that it is neither exclusively true nor exclusively false, but this can be interpreted in two different ways. If 'true' is represented by the Boolean value 1, and 'false' by 0, then the value ½ can be interpreted as 'both true and false', in which case the logic contains a glut and must be paraconsistent, or it can be interpreted as 'neither true nor false', in which case the logic has a gap and must be paracomplete. The models for these approaches are Graham Priest's 'logic of paradox'—*LP*—and Stephen Cole Kleene's 'logic of indeterminacy'—*K3*—respectively.[174] Reiterating what was said earlier, unless the principle of double negation equivalence is denied, a gap and a glut will collapse into each other. In a logic *with* this principle, the paracomplete theorist cannot assert that the paradoxical statement is neither true nor false but can only *deny* it is either. Similarly, the paraconsistent theorist cannot deny that it is neither but must accept it as a gap as well as a glut.

Returning to Tarski's undefinability theorem, if we maintain classical logic, there cannot be a statement within a formal language that asserts its own truth, and there cannot be a statement which defines what it means for its sentences to be true or false. As such, there can be no way to express that what we prove is actually true, and so the concepts of proof and truth have move much further apart. This is a much more significant result than Gödel's theorems alone, for while the latter speak solely on the syntactical properties of certain formal systems—that we cannot produce certain statements from certain rules—Tarski's theorem talks about the semantic properties of formal languages more generally—that we cannot define a concept such as 'truth' without some false statements being contained in its definition. Without a hierarchy of metalanguages or an additional truth value, mathematical truth cannot be defined in mathematics, physical truth cannot be defined in physics, philosophical truth cannot be defined in philosophy, and so on for any formal language.

In conclusion, we have seen that within all of the most important notions of philosophy, we can derive the dialectical matrix. We usually appeal to logic in our claims that only one of these views can be right, as expressed by the position I call 'monoletheism'. However, given that we can produce the dialectical matrix not just for logical consequence, but also for truth itself, we cannot simply appeal to logic to justify our discriminations in philosophy. On the contrary, using monoletheism as the basis of discrimination between logics requires an assumption that monoletheism is also the conclusion of that discrimination, which is begging the question and a vicious circle. We are therefore forced to acknowledge that, in addition to the possibility that one thesis is true and its antithesis false, neither or both of them could be true as well.

Chapter Seventeen

Computation & Consciousness

The Anti-Mechanist Argument

Gödel's theorems are sometimes glossed as providing evidence for fantastical claims ranging from the limits of science to even the existence of God. Such claims are not taken seriously, for Gödel's theorems properly concern arithmetic and arithmetic alone. Nevertheless, mathematics is not just an artefact of abstract formal systems but is fundamental to intelligence on a physical level. We are mostly done with all technical discussion on formal systems at this point, but there are several important results in mathematics following Gödel that I must mention before progressing. These results reveal the incompleteness theorems as instances of a much more concrete limitation regarding the nature of mechanics.

First, there is the Church-Turing thesis, which showed that an arithmetical problem is decidable just in case it is mechanically computable; and second, there is Turing's reproduction of a self-referential paradox within the architecture of computation. The latter showed that there can be no effective method, and no Turing machine, for determining whether a program can solve any given problem or not. Turing's 'halting problem' takes the incompleteness theorems and conveys them in a concrete manner, where the undecidability of the Gödel sentence becomes the non-computability of the halting problem.

While the incompleteness theorems tell us nothing mystical about the world beyond mathematics, it is important to realise that they are really just a magnified expression of a much more general characteristic of self-referential systems. It is in the isomorphism between these instances that the essence of the limitation lies. Due to the isomorphism between abstract decision procedures in formal systems of arithmetic, and physical computational procedures in machines, a genuine philosophical implication of Gödel's result concerns what it means for the basic architecture of human intelligence.

Recall that the Gödel sentence of a given formal system, which asserts its own unprovability, can be seen to be true necessarily, since if it were false, it would be provably unprovable. The truth of the Gödel sentence can be established from outside of the theory it is produced in, but we do not *need* a metatheory to *know* that it is true—its truth is intuitively clear. Since a formal system is built of a collection of axioms that could be mechanised, equally any computer can be modelled as a formal system. If we consider that the human brain is at root a biological computer, so too could the brain be modelled mathematically, and so too would it have its own Gödel sentence.[175]

Now, the brain, at the level of its computations, is clearly consistent, and so it cannot be the case that the firing of neurons could produce its own Gödel sentence, on pain of contradiction. Nevertheless, we know that our brain's Gödel sentence is true, again, on pain of contradiction, and so immediately it appears that the human mind can produce something that the brain cannot. Moreover, the mind appears to be fortified with a fail-safe method of avoiding any negative consequence of self-referential loops. A mind cannot get stuck in a spiral of madness when considering the truth status of 'This statement is false'. A quick mind may see the self-reference from the outset, needing not to test the premise, and even the slowest of minds need only follow the loop once around before realising it takes us back to where we started.

There is no end to the loop of self-reference, but we provide it an end by understanding some deeper meaning. The question is, how should a machine do the same? If it is not programmed to stop searching for a conclusion, it may continue forever, and if it is programmed to stop, how soon should that be? What if a mistake was made the first time around? Clearly, any suitable model of the brain will surely have some mechanism in place to prevent itself from madness, but could it produce the answer: 'The liar is both true and false'? If not, then neither can the brain, at least at its lowest level, encode the notion of truth.

One conclusion that has been drawn from the seeming incongruence between human intelligence and computation is that it reveals the human mind to be non-mechanical. This idea in fact stems from Gödel himself, who went to some length in an attempt of proving it. A full discussion of Gödel's contention that "not all mathematical thinking is computational", and also that mind is separate from matter, is given by Hao Wang in *A Logical Journey: From Gödel to Philosophy*. Wang quotes Gödel:

> My incompleteness theorem makes it likely that mind is not mechanical, or else mind cannot understand its own mechanism. If my result is taken together with the rationalistic attitude which Hilbert had and which was not refuted by my results, then [we can infer] the sharp result that mind is not mechanical. This is so, because, if the mind were a machine, there would, contrary to this rationalistic attitude, exist number-theoretic questions undecidable for the human mind.[176]

Hilbert's rationalistic attitude is of course embodied by his formalism—that there are no mathematical problems mathematicians are incapable of solving. This was widely thought to have been defeated by the incompleteness theorems, though Gödel himself remained unconvinced, likely because he was unperturbed by the consequences of it being true in light of them. However, the Gödel sentence of a brain is not an undecidable statement for mathematics as a whole, but only for the system it exists in. As we shall see, there is the possibility that our intuitive production of the Gödel sentence of our brain is similar to a metatheoretical proof. Gödel is talking about *genuinely* undecidable problems, like the continuum hypothesis; if there is some way or another for us to work out solutions to these undecidable problems, perhaps using non-classical logic, *then* the mind is not mechanical.

The issue regarding the brain's Gödel sentence was notoriously raised in J.R. Lucas' 1961 essay entitled 'Minds, Machines and Gödel', and again later by the Nobel Prize winning physicist Roger Penrose. Their argument is quite simple: any computer that is sufficient for modelling the mind will contain at least one true sentence—a Gödel sentence—which it will be unable to prove; since the human mind can see that this sentence is true, the mind must have some capacity the brain lacks.[177]

The above argument has not been particularly well-received by the philosophical community, for while the conclusion does indeed follow from a reasonable set of premises, it is possible that one of these premises is false. Moreover, denying mechanism is commonly seen to be a very undesirable result. One premise we might doubt is that the human mind is consistent, for while we cannot hold beliefs of the form A and not-A, we can at least hold beliefs that are incompatible. Since the incompleteness theorems only apply to consistent systems, it has been argued that they do not apply to the mind insofar as the mind is inconsistent.[178]

The precise sense in which we are inconsistent is an important point, however, for we certainly aren't inconsistent wittingly, and our mistakes are often intelligent mistakes that are corrected once made known. Again, Aristotle thought it impossible to simultaneously believe the same thing can both be and not be, but this would not prevent us from holding disconnected beliefs that are conflicting on closer inspection. Moreover, it is a common view that there is a one-to-one correspondence between neurological and psychological phenomena, yet we cannot claim that we are inconsistent at the level of our neurons. It must rather be the case that our inconsistencies are riding on something—the brain—which is itself consistent.

Lucas responds by arguing that the mind cannot be inconsistent, despite its inconsistencies, for there would be no way to program the fallacies of cognition into a formal system without rules to determine which inconsistencies are to be permitted and which to be denied. The acceptance of mistakes at random cannot suffice as an adequate model of the mind, and since such rules would be susceptible to Gödel's theorems, if the mind is truly inconsistent, it must be inconsistent at its lowest level.[179] Of course, if the mind is mechanical, it is surely many orders of magnitude more complex than any formal system we have so far devised, and we certainly cannot be sure of how it could or couldn't work.

In regard to this specific issue, I think it most likely that the mind cannot be described as a consistent system and does not fall victim to the incompleteness theorems. I emphasise that this does not entail that the mind is mechanical, however, for it could still be inconsistent *and* non-mechanical. Either way, the anti-mechanist argument is important to our discussion. At least, it implies that the mind is inconsistent *or* non-mechanical, and therefore, crucially, that the mind is not a consistent machine. This raises the equally important question as to how such a system can be grounded in the neural network of the brain, which at root *is* consistent and mechanical. As an aside, let us acknowledge the irony involved in the idea that an inconsistent thing could reason that itself is inconsistent, without doubting that the method used to get there isn't inconsistent too!

Brain vs. Mind

The majority of the discussion on anti-mechanist arguments has focussed on Gödel's theorems, but Tarski's is of equal importance. For the mind to contain its own truth predicate, it must be inconsistent or non-mechanical, and since the neural structure of the brain is both consistent *and* mechanical, *it* cannot produce the notion of truth. While the anti-mechanist argument does not conclusively demonstrate that there is some facet of the mind that defies mechanical explanation, it does lead us to this image of a mind composed of two parts—of a neurological computer, which is the consistent formal system, and an inconsistent program, in which we find semantic notions such as truth.

There must be a fundamental difference between these two parts of human cognition, and it is likely that their underlying logics are distinct. If this is so, it would seem that there is a complementary relationship between the inane syntactical procedures of the brain, and the intelligent appearance of semantic meaning in the mind. Our raw intellect must be substantiated by our neurology but become obscured at the level of mind, as mechanistic operations are symbolically sublated together. Conversely, our conceptual intelligence must be facilitated by our psychology, but is deconstructed at the level of the brain. With this consideration in mind, we have made our way to the central thesis of the American cognitive scientist Douglas Hofstadter's classic *Gödel, Escher, Bach*. As Hofstadter states, one of the main aims of this work was to show that:

> Every aspect of thinking can be viewed as a high-level description of a system which, on a low level, is governed by simple, even formal, rules. The "system", of course, is a brain...The image is that of a formal system underlying an "informal system"—a system which can, for instance, make puns, discover number patterns, forget names, make awful blunders in chess, and so forth. This is what one sees from the outside: its informal, overt, software level. By contrast, it has a formal, hidden, hardware level (or "substrate") which is a formidably complex mechanism that makes transitions

> from state to state according to definite rules physically embodied in it, and according to the input of signals which impinge on it.[180]

Hofstadter describes the brain as an incredibly complex yet fundamentally logical mathematical object, which can only be understood by "chunking it on higher and higher levels, and thereby losing some precision at each step". At the level of neurons, there is no intelligent interpretation of why which and when neurons fire, while "on the top level, there emerges a meaningful interpretation", for it is there that the mind is a reflection of our conscious experience of the world. Hofstadter's response to the anti-mechanist argument of Lucas is that while Gödel's theorem can be applied to the neural structure of the brain, just as it can be applied to any consistent formal system, *that* is not the level of our intelligence, nor of our knowledge of the truth of the Gödel sentence. While the brain cannot prove its own Gödel sentence or define its own concept of truth, the mind can.

If the brain, being a consistent mechanism, contained its own truth-predicate, then it could encode the liar paradox into the firing of its neurons. The brain's attempt to determine the truth value of the paradox would then "physically disrupt the coding of the sentence", which is "tantamount to trying to force a record player to play its self-breaking record". Indeed, Hofstadter believes that the standard, natural language version of the paradox "involves abandoning the notion that a brain could ever provide a fully accurate representation for the notion of truth...that a total model of truth is impossible for quite *physical* reasons: namely, such a modelling would require physically incompatible events to occur in a brain".[181]

These two levels of cognition at least superficially seem to correlate with the dialectical matrix. That is, the symbolic software level can be associated with the subjectivist quadrant, and the neurological hardware level can be associated with the objectivist one. We have seen already from the dialectical matrix for the notion of truth that these quadrants are associated with paraconsistent and paracomplete logic respectively. If the mind contains contradictory symbols which have no equivalent representation in our neurons, then the inconsistent mind and the incomplete brain fit neatly into our matrix. We can complete this image by moving on to consider Hofstadter's central thesis, and a third component of the mind-body relationship, which expresses a tangle between the two levels just as the superjectivist perspective expresses a tangle between the subjective and the objective generally.

Strange Loops

The characterising mark of the paradoxical behaviour which arises in the presence of self-reference, is that an action, activity, or proclamation is taken and transformed into its negation. Self-reference in language and logic arises when the consequence of a statement or proposition is recycled and used as its own premise. It is self-containment which generates these loops, for while we think of containers as being larger than their contents, a self-container is identical to that which it contains. Contradiction occurs when the distinction between containing and being contained is so blurred as to become indistinguishable; it is nothing to do with the actual domain in which it occurs.

Accordingly, we find similar paradoxical structures occurring in a variety of different forms—with sets, with words, with sentences, predicates, and even with barbers who shave only non-self-shaving men. The paradox is always essentially the same, for it is really a paradox *schema* that expresses an inescapable facet of being a sufficiently powerful subject. We characterise the paradox in the negative because its consequences are more striking, but positive self-reference conveys the same non-Boolean characteristic. For example, the word 'non-self-descriptive' describes itself if it doesn't and doesn't if it does, but the word 'self-descriptive' describes itself if it does and doesn't if it doesn't—both are consistent, and both can be true.

Any system with an unrestricted capacity for the extension of its action—whether that action be predication, containment, perception, or any kind of reference—will produce a feedback loop whenever the function of the subject extends over itself. When self-reference occurs, a single entity is acting both as subject and object, and so self-reference embodies a coalescence between the two fundamental aspects of our experience—subjectivity and objectivity. We cannot use any computational procedure to make conclusions about self-referential structures, because any system of instructions will continue until it is instructed externally to cease. Yet, the human mind has no need to follow the recession of feedback in determining whether it ends, for it is clear to see that 'This statement is false' cannot be solely true or false, and it is equally obvious that the sentence which asserts its own unprovability is true.

Our minds are a constant cacophony of ideas, beliefs, sensations, feelings, and perceptions, all constructed of mental objects temporarily filling the space of our awareness. They rise and fall, furnishing our experience, but always relying on that which is still, which is the space beneath them. This space is nothing else but a perpetually self-referencing awareness of oneself, whereby we are not only conscious, but are conscious of being conscious, and this is not disturbing at all. The space is easily ignored when more interesting things are imposed upon it, like marks on a blank canvas, but it is always present throughout our thought. It is the one thing we are never *not* confronted with, and so it is the most basic, and most obvious, element of experience.

At the same time, when actively recognised, self-reference is the most fascinating of all phenomena. Its perpetual self-recycling motion seems to confer a sense of intelligence; it *is* the foundation of *our* intelligence. Distinctions become indistinct in self-reference; unity, duality, and infinity all play together in the endless cycling of information. Self-reference is dualistic in that it involves that which is referenced and that which references, unitary because these two aspects are played by a single thing, and infinitary because the process of self-reflection is inherently endless. The feedback between subject and object generates motion without an external cause, which, in examples of feedback loops in physical systems, exudes the appearance of an autonomous will.

Hofstadter recounts numerous instances of self-referential systems from widely diverse disciplines, a particularly illustrative example of which occurs in video feedback, where a camera is pointed towards an image of what it is capturing. The resultant image warps and moves, evolving as it self-recycles without the need of external stimuli. The only change occurs to the change itself, self-modifying its own self-modification in a kind of spontaneous metamorphosis. The association with vitality and intelligence is made explicit in the self-reference of biological systems—in the regulation of neurochemicals, hormones, and other homeostatic functions. Hofstadter develops an analogy between Gödel's theorems themselves and the minutest levels of genetic replication—between mathematical formulas and DNA sequences, between Gödel coding into unique numbers and genetic coding into proteins, and between formal self-reference and self-reproduction in the cell.[182] Hofstadter's goal, however, is to reveal a much greater role for self-reference in the organism—the production of consciousness in man.

In talking of the mind as a series of levels above the brain, any physicalistic perspective must maintain that the brain is the most powerful level of all, for that is the level that does all the work. The mind is subject to continual self-modification, being software, but the brain is locked as an inviolate structure, as hardware. Despite the fact that the brain does all the work, however, it must be the mind that acts as its metatheory, and not the other way around. That is because the mind has access to ideas that are cut off from the brain necessarily—things like Gödel sentences and the notion of truth. Hofstadter writes:

> Gödel's proof suggests—though by no means does it prove!—that there could be some high-level way of viewing the mind/brain, involving concepts which do not appear on lower levels, and that this level might have explanatory power that does not exist—not even in principle—on lower levels.[183]

As stimuli are chunked onto higher levels, developing into symbols that model the world, eventually symbols emerge that model their own source, and the levels themselves are given representation in the mind. Since all of these levels are, in a real sense, 'squashed' onto the neural structure of the brain, they are at once many and one—many levels of software, but always one level of hardware, which encodes the entire hierarchy. It is this that leads Hofstadter to the terms 'tangled hierarchy' and 'strange loop'.

Hofstadter sees a connection between our ability to understand the meaning of the Gödel sentence, without needing to deduce it from some axioms, and our unshakable inclination that high level constructs such as beliefs, desires, and ideas 'command the ship', so to speak. He writes:

> My belief is that the explanations of "emergent" phenomena in our brains—for instance, ideas, hopes, images, analogies, and finally consciousness and free will—are based on a kind of Strange Loop, an interaction between levels in which the top level reaches back down towards the bottom level and influences it, while at the same time being itself determined by the bottom level. In other words, a self-

> reinforcing "resonance" between different levels—quite like the Henkin sentence which, by merely asserting its own provability, actually becomes provable. The self comes into being at the moment it has the power to reflect itself....This act of translation from low-level physical hardware to high-level psychological software is analogous to the translation of number-theoretical statements into metamathematical statements. Recall that the level-crossing which takes place at this exact translation point is what creates Gödel's incompleteness and the self-proving character of a Henkin sentence. I postulate that a similar level-crossing is what creates our nearly unanalysable feelings of self.[184]

Note that a Henkin sentence is comparable to a Gödel sentence without the negation, for its own structure provides its own derivation. Again, the analogy, which may be more than an analogy, is between low-level physical hardware (the brain) and number-theoretical statements (an incomplete system), and between high-level psychological software (the mind) and metamathematical statements (an inconsistent system). The strangeness of the strange loop emerges because the mental metasystem provides meaning to the physical system, while also being embodied within it. Thus, for one to affect the other is also for it to affect itself, and thus its effect on the other, which again affects itself.

To be clear, Hofstadter does not extend the analogy to non-classical logics—that addition is my own—and much of the work on them has been done since the first publication of *Gödel, Escher, Bach*. Hofstadter is also operating within the purview of a physicalist ontology. He is somewhat of a unique physicalist because he accepts the reality of free will and consciousness as at least physically inexplicable, and he also accepts downward causation between the mind and brain, but he is a physicalist nonetheless—specifically, I think, a *weak emergentist*.

In regard to these matters, some form of ontological reductionism is most often our default position, and for Lucas to even be able to disprove mechanism, we would have to assume it to begin with. The perspective is an objectivistic one, for there is one level that can never be modified by the higher-levels, and that is the laws of physics which determine the rules of computation for the hardware. The subjectivist might argue, however, that there is also a level on the other end of the hierarchy that is invulnerable to the mechanism, and that is the immutability of consciously recognised meanings. Moreover, the same self-asserting character of consciousness, which Hofstadter presents from a physicalistic perspective, has also been presented from an idealistic one. We find it prominently in Hegel, but also particularly in the work of Johann Fichte.

Fichte was a contemporary of Kant who expanded on the latter's transcendental idealism. He was also the first German idealist in the lineage connecting Kant to Hegel, being a teacher of Friedrich Schelling, to whom Hegel had a long and interesting relation. Fichte sought to ground philosophy in an indubitable first principle, which for him was the logical law of identity, *A=A*. He saw this law as the logical expression of self-position, and the union of an *act* and a *fact*. Fichte identifies self-position as the 'I', so that the possibility of the self positing its own existence—an act—is precisely the necessity of its own existence—a fact. He writes:

> The self's own positing is thus its own pure activity. The self posits itself, and by virtue of this mere self-assertion it exists; and conversely, the self exists and posits its own existence by virtue of merely existing. It is at once the agent and the product of action; the active, and what the activity brings about; action and deed are one and the same, and hence the 'I am' expresses an Act, and the only one possible, as will inevitably appear from the Science of Knowledge as a whole.[185]

Whether interpreted from a physicalistic perspective or not, what we can take from Hofstadter's thesis is that there is enough difference between the brain and the mind to call them different things. There is no sharp or obvious distinction between them, and their interactional tangledness can be seen both to separate and unify, but there is clear complementarity found in their relation, for each provides something the other cannot. Consistency reduces as we go up the levels, and completeness reduces as we go down them; the brain is purely syntactical, while the mind is concerned with semantics.

We can now begin to think of human cognition as a composite of three components that conform to the dialectical matrix. The brain, and the body by extension, is the objective aspect, which is consistent but incomplete; the mind is the subjective aspect, being complete but inconsistent; and consciousness itself is the superjective aspect, expressing a feedback loop between the mind and brain, and in which the subject and object are identified. It is also the latter which is syntheoretic with the paradoxical components of other areas in philosophy, and this shall be the topic of the following chapter.

Part IV

On the Shape of Knowledge

Chapter Eighteen

Philosophy of Self

Metaphilosophy

Philosophy is not like the scientific disciplines, for the nature of science makes no odds to the scientist's ability to undertake their task. With philosophy, however, there can be no hope of success in absence of understanding precisely what philosophy is. Even in the psychological sciences, we are aware of the object of our inquiry, but in philosophy we have not this simplest of facts. We do not know what we study in philosophy, and that is precisely why there is philosophy, and why it is not yet a science.

Of course, if philosophy is to talk about anything at all then it should talk about that which is actual, but to do this we need to know already what the actual is, and for this we need philosophy. All we can say, then, is that philosophy talks about what *could be* actual, and in this sense, it is limited only by what philosophers can conceive of as being actual. Philosophers study their own intellectual instincts, and what makes a good philosopher is an ability to see these instincts clearly—from where, why, and how they arise—and then mould them into something intelligible to others. What makes a good philosopher is a dissatisfaction with confusion; if we can formulate ideas that make sense to ourselves, then we have done our job justly; if we can formulate ideas that make sense to others, then it seems we have captured something real.

The object of philosophy is ineluctably oneself. We are investigating our own ability to investigate, and any progress in this ability engenders changes in the very process by which one thinks, which changes the result of one's thought, which changes the process of thinking again. This is something I have experienced strongly throughout the course of completing this work, as my passion, which was once unconcerned with rigour, has been transformed in an attempt to convey myself appropriately for public dissertation. In discourse, the emphasis is continually flipped from what it is one wants to say to how it is justly said, and these things endlessly influence each other.

One thing that is constant in metaphysical philosophy, which is not in natural science, is our intention to talk of the world as a whole, and so the theorist is always a component of their own theories. The metaphysician describes a world containing themselves describing a world. The phenomenologist describes experience from the perspective of experience. Any philosophical theory is in part a theory of the nature of theories, and the mental processes by which we access those theories. If a theory does not acknowledge and account for the full extent of its universality, then it can never explain how itself came to be. This work, for instance, can be nothing more than a work about myself; no matter how much I try to avoid self-reference, everything is personal in philosophy.

Self-reference is the fuel of philosophy. We believe about how we believe; we reason about how we reason; and there is no philosophy which is not metaphilosophy. The difference between a fanciful belief and a stringent philosophical conviction is made by how far down the path of self-reference we go, before bottoming out on some disbelief or doubt or another. To believe fully, we must believe we are justified in belief, and that our belief in this belief is justified, and so forth. We cannot truly convey our basic, instinctual, and temperamental beliefs without involving others that are themselves about beliefs. It is always our basic beliefs that drive us, and the others are mere artistry and rhetoric. The question arises: is a belief in one's basic level beliefs a basic belief or a meta-belief?

The paradox of self-reference emerges in all philosophic disciplines for any understanding must also understand the nature of understanding. Epistemology must inquire into the means by which we can ascertain the proper route to knowledge; ontology must define existence as a whole in terms of a fragment of itself; and axiology must evaluate the capacity to judge value. There are inquiries into the nature of the world, but there are also inquiries into the nature of our inquiries. Does the inquiry into our inquiries into nature inquire into itself? Is metaphilosophy a subject of philosophy, or is it something else?

The universal extension of philosophic notions evokes paradoxes that can be addressed from complementary perspectives, much like the universal set in logic. Though we may construe our thought as to ignore the paradox, any theory universal enough to explain the full range of phenomena will also have to explain the nature of paradox. We have already seen in Part II that within each of the major disciplines of philosophy, there arises a self-contradictory aspect of our theorisation, which involves the co-instantiation of complementary properties. We shall now return to this aspect of the dialectical matrix, and we shall see that it is fundamentally related to the strange loop which is our own self-consciousness.

The paradoxical component of philosophy is analogous to the strangeness of self-consciousness, for just as the latter expresses a union between being a subject and an object of perception, the superjective expresses a union between the contents of subjectivity and objectivity generally. Philosophy is a magnification and systematisation of conscious thought, and it is integral to conscious life whether it occurs in the academy or not. There is no consciousness without philosophising, just as there is no philosophising without consciousness, and we are all philosophers at heart. If self-referential activity is central to the basic mechanism of thought, then it shall be central in any formalism thereof.

The Analytic A Posteriori Revisited

We are now ready to return to that aspect of the dialectical matrix which gives rise to self-contradictory notions in all areas of philosophy and provides it with a newfound description. Superjectivism expresses a confluence of the subjective, which is characterised by immutability, and the objective, which is characterised by determinateness.

	Indeterminate	Determinate
Mutable	**Abjectivism** Synthetic A Priori Abstract Particular Objective Relative Authoritarian Individual	**Objectivism** Synthetic A Posteriori Concrete Particular Subjective Relative Libertarian Individual
Immutable	**Subjectivism** Analytic A Priori Abstract Universal Objective Absolute Authoritarian Collective	**Superjectivism** Analytic A Posteriori Concrete Universal Subjective Absolute Libertarian Collective

In epistemology, this confluence reveals itself as the *analytic a posteriori*; in ontology, as the *concrete universal*; in axiology, as the *subjective absolute*; and in political theory, as *libertarian collectivism*. All of these notions express a conflict, for each of their components preclude the possibility of the other when considering the objects of our perception. For this reason, each has been denied meaning in philosophy, and save for a few unique cases, seldom have they been proclaimed as the central features of normative theories of the world. Despite this, we shall now see that these concepts do have meaning, and they are in fact necessary components of the philosophic endeavour, falling predictably within the structure we have been uncovering throughout this work.

To begin, we have the *analytic a posteriori,* which refers to propositions that are justified via experience of the world—a posteriori—yet also by the logical relations of the concepts they contain—analytic. Under most interpretations, these two terms conflict with each other and so we posit their mutual exclusivity. Any truth that is justified analytically does not need to be experienced to be known to be true, and so an a posteriori confirmation of an idea already known will fail to be a condition of its truth—it will be *a priori.* To boot, Kant used the term 'analytic' simply to refer to logical knowledge, and the term 'a posteriori' to refer to the empirical. He did this to distinguish the *synthetic a priori,* which was his central aim to prove. Insofar as he had not considered analytic a posteriori knowledge to be possible, it made sense to him to involve 'analytic' with 'a priori' and 'a posteriori' with 'synthetic'.

Nevertheless, Stephen Palmquist, who is one of the world's leading scholars on Kant, has argued that Kant was mistaken in this conflation, and that the analytic a posteriori constitutes a genuine class of knowledge:

> The impossibility of analytic a posteriori knowledge is generally considered to be 'quite evident' [P5:182-3]: indeed, it is a nonsensical contradiction in terms for those who equate 'analytic' and 'a priori' [see Ap. IV]. Even though Kant argues against those who identify analyticity and apriority [e.g., in Kt1:1-10], he joins them in dismissing this class of knowledge with only a brief explanation: 'it would be absurd to found an analytic judgment on experience. Since, in forming the judgment, I must not go outside my concept, there is no need to appeal to the testimony of experience in its support' [Kt1:11; cf. Kt2:268 and Kt4:12]. There are, however, a few theorists who do regard the analytic a posteriori as providing the best

> description of certain types of knowledge. Notwithstanding Kant's lack of concern for this class of knowledge, I shall argue...that certain aspects of his philosophy can best be understood by reinterpreting them in terms of the analytic a posteriori. At this point, though, it will suffice to say that we should expect such knowledge, if it is possible, to have its validity grounded in some way in experience (a posteriori), and yet also to proceed by making inferences solely on the (analytic) basis of an application of the laws of logic to the concepts or propositions involved.[186]

The main barrier to analytic a posteriori knowledge is that it would require the innate justification of an object to be fundamentally involved with any experience of it. In other words, the fact of its truth, and the act of knowing it, cannot be disconnected, so that there can be no possibility of it being known *prior* to experience. It is now clear that there is one kind of entity that fits this description entirely, and which is known via precisely the analytic a posteriori. This entity is, of course, consciousness itself.

We conceive the necessity of our being conscious analytically for consciousness is required for the very possibility of its conception, yet it is also true that never do we *not* recognise our consciousness in absence of a posteriori experience of it. We assert our own awareness by means of Cartesian *cogito* type statements, such as 'I am conscious of being conscious, therefore consciousness exists', or 'I have consciousness, therefore I exist'. The propositional form of the statement is even unnecessary, since consciousness and existence are transparent predicates, so that we may simply assert 'I exist' or 'I am conscious' just the same.

As it happens, Palmquist argues that Descartes' cogito: 'I think, therefore I am', only makes sense if it is taken as analytic a posteriori. I find Fichte's 'I am I' to be an even better candidate, for he directly identified the *act* of self-assertion with the *fact* of self-being, though he took this further as to imply an act of self-creation. Palmquist also considers some arguments made by Saul Kripke and Philip Kitcher regarding the 'contingent a priori' status of statements such as 'I exist', 'I have some beliefs', and 'there are thoughts'. On his view, these too should be considered analytic a posteriori. He explains:

> Reference to the act of 'thinking about the issue' and forming a 'belief' imply that the statement 'I exist' is not a priori, but *a posteriori,* since the relevant aspect of these acts is that they are *experienced*. Likewise, the fact that this belief is implied as 'a product of my reflection' makes it *analytic*.[187]

The situation is perfectly natural to somebody working in syntheorology, for the strange loop between subject and object, which is conveyed in our being conscious of being conscious, reflects the equivalence between 'analytic' and 'a posteriori' in this particular case. Pure reason is just experience when it concerns awareness of the self, just as experience is pure reason; it is only through the separation of subject and object that the two can be seen as mutually exclusive. The analytic a posteriori is the point of fusion between the subjective realm of the analytic a priori and the objective domain of the synthetic a posteriori, just as self-consciousness is the bridge between the abstract domain of mind and the concrete external world.

The Concrete Universal Revisited

Next up is the contradictory aspect of ontology, which is the *concrete universal*. This refers to an object which is spatiotemporal while also being a category that objects share in common. The concrete universal is largely overlooked in philosophy, for all properties appear abstract, while all concreta are particularised. The mutability of the physical enables its determinability, but the concrete universal must be determinable without being changed by the act of being determined. It must be grasped of our sense of time and space but not limited to anything *in* them. It is the universal of universality, in which all things participate, and which participates in all things. As the analytic a posteriori is the *knowledge* of Self, so the concrete universal is the *being* of Self. It is the very notion of being, which is the possibility of its recognition, and which is the presupposition of the knowledge of all other things.

Knowledge and being are thus united in the Self, for the subjective recognition of 'I' is the substance of its objective reality, which is in turn a condition of its recognition. It is concrete in the sense that it is the essence of concreteness, and it participates in its instances just as it does itself. On one hand, the Self can never be pinned down, for the recursion between the knower and the known presents an infinite aspect to the Self that cannot be captured by thought. On the other hand, all that is known is known *to* the Self, and there is no knowledge of anything, which is not knowledge of the acquisition of that knowledge. The concrete universal *is* self-reference, and all those things that are what they are via participating *in* themselves; it is the Knowledge of Knowledge, the Truth of Truth, the Value of Value, and the Being of Being.

The concrete universal is best known through the work of Georg Hegel, in which it possesses an important position. For Hegel, the concrete universal is the universal of *individuality*, for individuals are united by the concrete universal as their substance, and the concrete universal is individual just in case it is particularised *as* individuals. While the instances of the abstract universal are merely examples of that universal, the concrete universal is *established* by its instances, for if all individuals were the same, they would not instantiate individuality. This self-participatory function of the concrete universal is lacking in the abstract. The abstract form of 'redness' is not itself red, for it is merely the idea of redness; the *concrete* idea is the Idea of Idea itself, or for Hegel the Concept of the Concept, and which is precisely self-consciousness. Hegel writes:

> The concept, when it has developed into a *concrete existence* that is itself free, is none other than the 'I' or pure self-consciousness. True, I have concepts, that is, determinate concepts; but the I is the pure concept itself, the concept that has come into *determinate existence*.

For Hegel, 'the Concept' reveals itself in consciousness as one learns what it means to be an individual. As one perceives the consciousness of others as distinct from oneself, and as one recognises oneself as the object of the consciousness of others, then one comes to see one's own self-consciousness as dependent on, and inseparable from, the consciousness of others. This evolution is underscored by a dialectical process in which the subject emerges as the object of its own thought, coming into a self-conscious unity with itself, and eventually merging completely with the objective world.

As a final point of note, Hegel speaks of our sense-certainty of the 'Here' and the 'Now' as something universal, and this is related to our recognition of self through the extension of spatiotemporality. Each 'here' and 'now' is particular, but our recognition of the 'here' and 'now' is only understood in regard to a multiplicity of 'heres' and 'nows', which are united in their 'here-ness' and 'now-ness' as universal. Thus, any fragmented experience of the spatiotemporal carries with it and substantiates all possible experience of the same, not merely as a whole with parts, but as its essence. Robert Stern explains:

> The fact that each 'now' and 'here' is always divisible into further 'nows' and 'heres' means that sense-certainty cannot claim access to just such a unique individual in its experience

of a temporal or spatial moment. Thus, even when it points and says 'now' or 'here', it is conscious of many instances of the same kind, and thus individuals that share the same property or universal (the property of being 'now' or 'here').[188]

The Subjective Absolute Revisited

The third element of the superjectivist perspective concerns axiology, and the notion that value is *subjective*—existing dependently on the mind—and *absolute*—applicable regardless of circumstance. The subjective absolute, just like the previous two concepts, represents a value system that is, ideologically, both subjectivistic *and* objectivistic. It is subjectivistic in its absoluteness, since, according to the subjectivist perspective, values are *inherently* meaningful, and therefore actions are right or wrong; it is objectivistic in its anti-realism, for on the objectivist perspective, abstractions do not exist.

In regard to normative ethics and the prescription of moral action, subjective absolutism leads us blindly, if not into circularity, at least into ignorance of proper value judgements. If we argue that the good and just depends on the minds of individuals, but also that individuals are bound to disagree, then it seems we shall be forced into relativism. If we deny this and assert that there are principles that ought never be violated, then it seems we are forced to accept that value is independent from the mind. There is thus a kind of self-negation that emerges in the normative subjective absolutist, whereby one's subjectivism denies one's absolutism, and one's absolutism denies one's subjectivism.

This self-negation is not undeniable, however, for there is a way in which the moral code of individuals might be subjective while nevertheless always arriving at the same conclusions. Our ability to arrive at proper value judgements may not be found in the perception of objective features of the world but may be present inherently in our own psychic structure, and our own capacity as moral agents. This line of thinking has a rich history and is central to any claim of the reliability of moral sense and reason. It is found for instance in the various formulations of the Golden Rule, and later in Kant's categorical imperative, which, again, dictates: "Act only in accordance to that maxim through which you can at the same time will that it should become a universal law."[189]

Kant is often presented as a moral realist due to his universalism, which is sometimes used as a synonym for moral objectivism, though the present work uses the latter as a synonym for moral *realism* only. It is unfortunate that our terminology in moral philosophy can become so easily muddled, not just in the conflation of the two dimensions of the dialectical matrix, but also in the fact that 'subjectivism' is a position of the objectivist syntheoretic ideology and vice versa. Nevertheless, we can easily cut through the confusion, for it is in part created by failing to acknowledge that subjectivism and absolutism are indeed compatible.

American Kant scholar Frederick Rauscher argues that Kant's ethics should be taken as anti-realist in the sense that it holds that "all of the moral characteristics of the world are dependent upon the human mind".[190] He highlights the mistake in equating realism with absolutism and explains how Kant's *anti-realist* position is compatible with the idea that moral truth applies universally. Rauscher refers to John Rawl's position that Kant's categorical imperative expresses an anti-realist constructive principle, in the sense of being a method for determining which maxims qualify for the status of moral law. He writes:

> These are said to be constructed because they do not reflect any prior moral order. The categorical imperative procedure itself is not the result of construction but rather "laid out" on the "basis [of] the conception of free and equal persons

> as reasonable and rational, a conception that is mirrored in the procedure" and "elicited from our moral experience." Thus, practical activity by agents who view themselves with a resulting collective self-conception provides the basis for a procedure that in turn provides the content for morality. On this view Kant is seen as a moral anti-realist because morality is not independent of the practice and self-conception of certain types of beings.[191]

Rauscher agrees with Rawl's position, though he contends that, it is "an internal experience peculiar to human-like rational beings" which is the foundation of moral law, that "the fact of reason consists of the categorical imperative as our consciousness of the moral law", and that "our consciousness of the moral law just is the categorical imperative".

The fact that moral knowledge should be accessible to human beings by the very fact of their reason implies that there is a *logical* connection between ethical behaviour and moral truth. To boot, Kant believed that we are held to the categorical imperative precisely because its denial would be illogical. If a universal law would prevent its own adherence when taken to a logical extreme, then that law must be false. It is precisely in the rational consistency of maxims that they may be validated by the categorical imperative. For instance, if everyone stole from others when it benefited them, then everything would be stolen, and nothing would be left to steal. It is therefore *illogical* that stealing could be moral. For Kant, this makes morality a matter of reason, even though it may coincide with emotion, and so just as a logical truth is rational everywhere and always, an ethical action is moral everywhere and always. This raises the question as to how the categorical imperative aligns with the epistemological component of superjectivism—the analytic a posteriori.

For Kant, moral action is grounded in the laws of duty just as logic is grounded in the laws of thought, and the irreducibility of the law makes the categorical imperative the primary axiom of our moral reason. Universal qualities such as goodness, justice, virtue, and value are then encoded by this law, and moral actions, insofar as they are moral, are contained logically within them. The subjective absolute then represents the coming together of principles and praxis, the subject and the object, and therefore also the analytic and a posteriori. Palmquist explains in considering the relation between logical and moral law:

> In both cases the law is analytic in relation to other laws in its system because it can be used to test the validity of such subordinate laws, yet it cannot itself be verified by appealing to a higher law. The difference is that, whereas logical laws are necessary a priori for all thinking and are thereby equally applicable in principle to all experience...practical laws apply to what ought to be the case, a posteriori, in 'matters of conduct', and 'allow for conditions under which what should happen often does not'. Thus, for example, we call someone 'good' by judging the extent to which their behaviour, considered a posteriori, coincides analytically with the idea of 'perfect goodness'—that is, the extent to which their behaviour is, as it were, 'contained in' that idea of perfection.[192]

In regard to the coherence of the superjectivist perspective, moral properties such as the 'idea of perfection' are recursive in the sense that they apply to themselves. Such notions can be found pervasively in the work of Plato, such as when he says, for instance, "Justice is of the nature of the just," "Nothing can be pious if piety is not pious," and, of course, in the Form of the Good. That is, a Form is an absolute perfection, and 'goodness' is a relative perfection, so that the Form of the Good is the perfection of perfection.[193]

Recall that such self-reference is not the case for all universals, for we would not assert that 'redness' is itself red, nor that 'manhood' is male. The various characterisations of the concrete universal are *always* self-referential, for self-consciousness is only what it is in its being conscious of itself. David Ellerman claims that 'self-participation' or 'self-predication' is a defining characteristic of concrete universality, and that recognition of this is important for mathematical logic, where in set theory self-participation was eliminated via Russell's vicious circle principle. He refers to set theory as a theory of abstract universality, while category theory is a theory of the concrete.[194]

Finally, subjective absolutism can be characterised as ideal observer theory, which posits that moral truth is rational, but only to a perfectly rational mind. Self-reference is overtly apparent in the ideal observer, for their capacity for moral agency has been identified with moral truth. Value judgements are vindicated precisely in their belonging to an ideal observer, and the ideal observer is ideal precisely in their making of proper value judgements. We have here an alternate version of the Euthyphro dilemma, which exposes the circularity in another subjective absolutist meta-ethic called 'divine command theory'. That is, something is good just in case it is approved of by an ideal observer, and the ideal observer approves of that something just in case it is good. Together these create a circle, as Plato's Socrates elucidates.[195] Goodness for the ideal observer is therefore a strange loop, and also a self-fulfilling prophecy, for by the very act of judging something as good, good so it is.

The Libertarian Collective Revisited

The fourth and final component of the superjectivist ideology to consider, though there will be countless more, is *libertarian collectivism*, which strives for the best of both worlds in exalting both the autonomy of the individual and the moral worth of the collective. Libertarian collectivism conveys subjectivism in holding that economic equality should be immutable, and objectivism in holding that social liberty should be determinate. Individuality is central to libertarian collectivism for the only way to unite sameness and difference is by making individuality universal among individuals. Libertarian collectivism therefore expresses a Heraclitean unity of opposites, for just as constancy is found in the ubiquity of change, equality may be found in the ubiquity of diversity.

Libertarian collectivism is often seen as a contradiction of terms which conflates antithetically opposed political philosophies. In Chapter Ten, we discussed Isaiah Berlin's conservative approach to liberalism, which stems from the conflict between equality, liberty, and other social values, and the need to maintain balance between such values. Freedom must be restricted to sustain total equality, for there will be no protection against inequality when there is total freedom. In the same vein, Belgian political theorist Chantal Mouffe states that "liberal democracy results from the articulation of two logics which are incompatible"[196] and that "once the articulation of the two principles has been effectuated—even if in a precarious way—each of them changes the identity of the other".[197]

Mouffe refers to this strange loop between democracy and liberalism as 'the democratic paradox'. It is a paradox because "the tension between equality and liberty" can only be managed via "contingent hegemonic forms of stabilisation of their conflict",[198] which, as we saw in Chapter Ten, forces our politic away from libertarian collectivism and towards authoritarian individualism. She thus elucidates an alternative approach to managing the paradox, which we shall return to shortly.

The acceptance of the conflict between equality and liberty does not need to be one of compromise, of course, for we may simply abandon one that the other may be maximised. Collectivists are naturally inclined to concede to authority, just as libertarians are inclined to admit inequality, and the logical conclusion of any extreme one-sidedness is state socialism or anarcho-capitalism respectively. It is not inconceivable that such positions may be favoured with good reason, for if compromise should call for the collapse of both values, then perhaps the greatest possible degree of value occurs when society discriminates.

The major issue with such prejudice, however, is that liberty and equality cannot be fully disentangled, for both imply an inequality of power that can be realised in neither the illiberal egalitarian society nor the liberal hierarchical one. There is a core value that is shared in both ideals, and it is, perhaps paradoxically, both achieved and destroyed through discrimination. This unity of opposites in the political sphere is quite simply the *sovereignty of the individual*.

If individuals would do what is required to achieve equality by the force of their own will, then no authority will be required to coerce them into action. Such a system would be self-regulatory at all levels, expressing the self-participation of the very notion of governance. Indeed, the society whose citizens have become ideal observers does not need to demonstrate the rule of law, for the distinction between what is good for oneself and what is good for others will have been dissolved. Such a democracy could not be crafted; it could only come from an organic, self-legitimising process that is a product of human feeling. Of course, this is presently nothing more than a utopian fantasy, but the fact that it is a fantasy does not make it inconceivable, and insofar as it is conceivable, it is also possible, and should not be ruled out as an eventual destination for civil life.

There is a syntheoretic relationship between the sovereign individual and the ideal observer, for it is only through proper moral agency that collective individuality can flourish, and the two strands of political thought may come together. The self-governance, self-ownership, and self-determination of such an alignment reveals not only the sovereign individual, but the ideal society as a whole, to be a concrete universal. This idea is explicated in Hegel's *Philosophy of Right*, where he provides an account of the state as the modality of the realisation of absolute freedom, via the dialectical development of society and reason.

For Hegel, 'the State' emerges as a concrete universal as the particular, subjective wills of individual citizens come into alignment with the universal will of Reason and the State as a whole. This alignment only occurs at the level Hegel refers to as 'ethical life', which "is the concept of freedom which has become the existing world and the nature of self-consciousness".[199] The sovereignty of the individual, which is the common thread in the complementary values of liberty and equality, is thus expressed in Hegel's thought through the notion of will: "*Duty* and *right* coincide in this identity of the universal and the particular will, and in the ethical realm, a human being has rights in so far as he has duties, and duties in so far as he has rights."[200]

It would be disingenuous to claim, however, that Hegel's thought aligns perfectly with the superjectivist perspective. That is because Hegel's state is not anarchical, and while he emphasised self-governance, he also respected the role of a centralised authority. It was not until Karl Marx took Hegel's dialectical method, and applied it to the conflict of social classes, that libertarian collectivism would be associated with the political philosophy of communism, the final outcome of which would be a complete synthesis of liberty and equality. Marx stripped the dialectic of its spirit and metaphysical foundation, arguing through it that a communist revolution would be the inevitable outcome of the historical process. However, the dialectic without spirit is no longer a logical and teleological process with a set destination, but a practical one fuelled by a violent war of opposites. Hegel, on the contrary, was not a revolutionary; he was a moderate collectivist who was concerned with the realisation of freedom, but also respected traditional virtues.

Chantal Mouffe offers an alternative solution to the conflict of liberty and equality, which is to say, alternate to the pessimistic acceptance of compromise, and alternate to the dialectical solution offered by Marx. As she sees it, liberal democracy as traditionally conceived is flawed insofar as the overt antagonism between liberty and equality entails the discrimination and oppression of differing opinions. She envisions reframing the dialectic into one that is *agonistic,* wherein there is no expectation that each ideal should be willing to concede to the other. "*Antagonism* is a struggle between enemies, while *agonism* is a struggle between adversaries," and an adversary is "somebody whose ideas we combat but whose right to defend those ideas we do not put into question"—they are respected.[201]

Mouffe describes her position as 'agonistic pluralism', for it acknowledges a pluralism of values to exist as competitors. Rather than allowing their conflict to be admonished as something that should give rise to a victor, however, it is instead exalted as the major driving force for positive change, and so diverse points of views are encouraged to participate in the ongoing process to realise freedom. Mouffe believes that this process should be cooperative and *dialogical,* whereas Marx's vision is combative and dialectical. I think this is an important distinction, and one that is given a metaphysical foundation by paraphilosophy in Chapter Twenty-One. If antagonism and dialectic are the heralds of contradiction, then agonism and dialogic are those of complementarity.

The End of the Beginning

With the elucidation of the superjectivist perspective, we have now completed construction of the dialectical matrix. There is an aspect of our ability to conceive the world which is inescapably self-referential, and it presents itself to our reason as the conjunction of opposing properties. These properties oppose each other not because they are directly contradictory, but because they presuppose each other's negations. This presupposition is given authority from the relativisation of philosophical notions, but collapses within the Absolute, which is the concrete universal.

The conceptual schema for the duality of opinion in philosophy is encapsulated by the irreconcilability of immutability and determinateness, and it arises whenever the concept extends outside of itself. When we speak of knowledge, we speak of the knowledge of something that is not itself knowledge—the truth *of*; the value *of*—and so we never see beyond the duality to predicate of the thing-in-itself. The determination of concreta alters that which we determine, and the immutability of the universal precludes the determination thereof. Only when the subject of knowledge, being, value, or right, ceases in seeking an object outside itself, does the capacity for determination align itself fully with the immutable.

As such, the analytic a posteriori is *self*-knowledge, the concrete universal is *self*-being, the subjective absolute is *self*-value, and libertarian collectivism is *self*-governance. Ultimately, these things are all united in the absolute unity of the subject and object and cannot be relativised without their bifurcation. We experience this bifurcation as a conflict, yet the conflict also provides us with the means to understand ourselves, whereby we find the Self arising in the interdependence and reciprocity of the thesis and the antithesis.

To conclude this section, it will be fitting to return our attention to the one man who has had more to say on the nature of the Self than anyone else; that is, Carl Jung. Central to Jung's analytic psychology was the idea that the Self was fundamentally constituted by a unity of opposites. These opposites become differentiated through our experience of the world, aiding survival, providing clarity, and enabling discrimination between ideas and perspectives, all in the development of individuality. Psychological dysfunctions are remedied, and development facilitated, through the reconciliation of opposites that have been repressed. "The self is made manifest in the opposites and in the conflicts between them; it is a *coincidentia oppositorum*. Hence, the way to the self begins with conflict."[202] As we shall soon see, a similar thing can be said for making progress in philosophy.

Chapter Nineteen

Sameness & Difference

The Beginning of the End

We have now arrived at a point in this work where we can begin development of the foundations of a science of philosophy, which will be the task of the following two chapters. The success and explanatory power of such a science will not be a simple matter to establish in the currents of academia, and only the future will be able to judge it. What we can take for granted, however, is that syntheorology is a genuinely novel science in the sense of being an objective study of the structure of human belief and reason, as it pertains to the mental exercise.

Any science of philosophy, no matter its purview, must be grounded in *indubitable* philosophical facts. Insofar as we have no such resolutions to the various problems of philosophy, these facts can only concern the process of philosophical inquiry itself. It may not be known, for example, whether materialism or idealism are true as ontological theories, but *it is* known that materialism and idealism are possible perspectives to have, and that they form part of a uniform structure of possible solutions defined by the dialectical matrix. These *are* philosophical facts, when the scope of philosophy is properly interpreted, and they are facts concerning the relationship between 'theory', as an object of thought, and 'theorisation', as a subjective procedure. So conceived, syntheorology bridges the gap between philosophy and psychology, providing a science of the pathways and potentia of human reason.

Nevertheless, by the end of this work, I hope I can portray a little more than this and penetrate through towards genuine philosophical truth. To demonstrate a correspondence between the possibilities of reason and the actuality of nature would be the means of such a science, and this would require establishing the dialectical matrix as the fundamental ontic existent. At least, we have shown already that philosophers are talking *about* the dialectical matrix, which is to say, about the ideas which are its elements. Whether philosophers, in all their disagreements, are also talking about the actual still remains to be seen. We begin this inquiry by discussing the isomorphic structure of the dialectical matrix.

Self-Reference & Duality

We have seen that Gödel's and Tarski's theorems provide the impetus for theoretical dualities in the abstract domain of formal systems, and we have also seen that a similar duality can be ascribed of the concrete domain of cognitive processes. Hofstadter believes that the ability for self-reference by the mind suggests that there are higher levels of operation present in neural processing, and that certain semantic constructions have no syntactic equivalent at the lowest level of the brain. The 'hardware level' of neurology and the 'software level' of psychology clearly fall into the objectivist and subjectivist quadrants of the dialectical matrix respectively. We have also seen that paracomplete logic and verificationism are situated in the objectivist quadrant and are therefore syntheoretic with the brain's inability to construct Gödel or liar-like strings and their negations. Furthermore, paraconsistent logic and falsificationism are situated in the subjectivist quadrant, being syntheoretic with the mind's *ability* to comprehend these self-referential negations.

In systems that permit them, self-referential negations produce results that cannot be captured by any static snapshot of that system. Recall that the liar paradox is true *if* false, and false *if* true, so that its status it always flipping to its obverse. The conjunction of the complementary principles of excluded middle and explosion, as expressed in the superjectivist quadrant, also entails a feedback loop of explosion and collapse, which can only be expressed *in time*. Consciousness also belongs to the superjectivist quadrant, involves a feedback loop between the subject and object, and is also expressed in time. Abstract systems do not *have* time, of course, so any intellectual resolution to the paradox requires the introduction of gaps or gluts.

Whether to admit of gaps or gluts respectively might seem at first to be an uneven affair, yet every benefit or drawback of one is mirrored by an equivalent benefit or drawback of the other. This is because both modify the definition of what it means to assert and deny, for the symmetry between paracompleteness and paraconsistency is preserved purely because the classical symmetry between acceptance and rejection can be broken in two equal but opposite ways. Paracompleteness entails that not everything that should be rejected is the negation of something to be accepted, and paraconsistency entails that not all negations of things to be accepted should be rejected.

When the believer in gaps states 'it is not the case that values are objective', and the believer in gluts states 'it is the case that values are not objective', the two are not in agreement, the former's statement being negative and the latter's positive. If the gap-theorist accepts that the negation of an assertion is a denial but denies that the negation of a denial is an assertion, the glut-theorist does the converse. We may be inclined towards gaps or gluts respectively, if either at all, but their symmetry precludes this inclination from being one decided by logic.

The asymmetry between assertion and denial is not characterised in classical thought, where every denial is an assertion and vice versa, but is nevertheless intuitive and crops up in various guises within philosophy. We have positive and negative descriptions of values, which are not equivalent, and we have positive and negative forms of beliefs, which are not equivalent. Positive or strong atheism, for instance, asserts that deities do not exist, while negative or weak atheism merely *denies* that they exist. The conviction of the former is stronger, but they cannot ground their belief on verificationism, while the opposite is true for the agnostic. This reveals a predicament of philosophy: the more you want to criticise something, the less means you have to do so.

Logical Pluralism

Just four years prior to Gödel's discovery of the mutual exclusivity of completeness and consistency for certain formal systems, Werner Heisenberg had discovered a similar limitative result regarding the mutual exclusivity of the expression of properties like position and momentum in particle physics. Heisenberg's findings provided the foundation for Bohr's principle of complementarity, which he proclaimed as a general principle of knowledge expressed in fields beyond physics. The principle was also put to use in psychology, both by William James before Bohr, and Carl Jung after him; and now we have the dialectical matrix, which captures our philosophic ideations in the language of complementarity, as articulated by the relation between the paradoxical superjectivist perspective and the subjectivist and objectivist perspectives of which it seeks the fusion.

Prior to the 20th century, logicians were quite content with classical logic as the sole foundation of rational thought, but if we are to consider the possible complementarity within various aspects of our world, we must not assume that the same logical framework is uniform between complementary perspectives. Rather, Bohr's vision of complementarity as a generalised epistemological principle would imply that it penetrates beyond the phenomenal, and beyond the theoretical, to the very notion of truth itself, so that there is no one single definition for the notion of logical consequence.

The Brazilian mathematician Newton da Costa, who was one of the early developers of paraconsistent logic, believes that subjects with a universal purview may need to be approached in the same complementary manner that Bohr highlighted for quantum physics. Recall that, in the previous chapter, we discussed how such subjects, in particular philosophy itself, are capable of extending and predicating over themselves, giving rise to paradox. The irreducible theoretical duality that results implies, in da Costa's words, that "there are several and eventually non-equivalent ways of looking at it (perhaps some of them based on non-classical logics), each one being adequate from its particular perspective, and showing details which cannot be seen from the other points of view".[203] He also writes that "different 'perspectives' of a domain of science may demand for distinct logical apparatuses" and thus "there is no just one 'true logic', and distinct logical (so as mathematical and perhaps even physical) systems...may be useful to approach different aspects of a wide field of knowledge".[204]

The traditional monoletheic view that there *is* just one true logic, just as there is one true philosophy, and one mathematics, has been challenged by the paradoxes of self-reference. Logicians have been compelled to look beyond the classical in the search for the true logic, yet the idea that there is one true logic is itself logical only if there is one true logic, grounding monoletheism again in circularity. If the law of excluded middle fails, perhaps there are no true logics; if the law of non-contradiction fails, perhaps there are many; and if the dialectical matrix reveals complementarity within our axioms, perhaps logical consequence is not a straightforward relation between premise and conclusion, but a complex structure admitting of multiple perspectives. The charge against logical monism has been led over the past 20 years by Jc Beall and Greg Restall, who summarise:

> This is our manifesto on *logical pluralism*. We argue that the notion of logical consequence doesn't pin down *one* deductive consequence relation, but rather, there are many of them...We should not search for *One True Logic*, since there are *Many*.[205]

This pluralism can also be extended to the truth predicate, considering that there may be more than one valid approach to the conception of a true sentence. Again, both paraconsistent and paracomplete approaches succeed in providing a 'transparent' or 'deflationary' account of the notion of truth, whereby a true sentence is simply one that states what is so, contrasting classical accounts which rely on a hierarchy of meta-languages. Since both give a different definition of what it means to accept something as true, it is also intuitive that there are multiple valid accounts of the concept of truth. The paracomplete theorist has trouble asserting sentences being *neither* true nor false, and the paraconsistent theorist has trouble distinguishing sentences being true and *not* false. Beall argues that these limitations can be resolved if we accept that assertions of the form 'not(*A* or not-*A*)'—a gap—and '*A* and not-*A*'—a glut—rely on different truth predicates than 'normal' bivalent truths.[206]

Indeed, the dialectical matrix for the notion of truth provides a clear path to truth pluralism, suggesting that truth-predicates are actually two-dimensional entities rather than Boolean. In this way, there are four predicates relevant to the notion of truth, each formed of one element of a dichotomy, much like all other positions on the dialectical matrix. That is: 'not true and not false', 'true and not false', 'not true and false', 'true and false'. Four additional predicates can be achieved under the relevant forms of negation, whereby acceptance for the paraconsistent falsificationist is simply being non-false, and rejection for the paracomplete verificationist is simply being non-true.

	Non-true	True
Non-false	Non-true & Non-false Sceptical	True & Non-false Verificationist
False	Non-true & False Falsificationist	True & False

Incidentally, but by no means accidentally, this structure of predicates mirrors both the Greek *tetralemma*, and the Buddhist *catuṣkoṭi*. There is also a non-classical four-valued logic called 'First Degree Entailment', to which Graham Priest has applied the status predicates of the *catuṣkoṭi* and so shown them to share a semantics.[207]

Everything & Nothing

The dialectical matrix characterises Truth as a concrete universal, expressing the complementary yet paradoxical union of truth and falsity, and which I distinguish by using capitalisation. Note that 'truth', being itself true, conforms to the self-referential nature of the concrete universal, just as 'individuality' is individual, 'value' is valuable, and 'self-consciousness' is self-conscious. Truth as a concrete universal, being a part of the superjectivist perspective, is therefore related to the glut-predicate—'both true and false'—and expresses a conceptual nonduality between truth and falsity. It is also related to analytic a posteriori self-knowledge, as the union of the empirical and the logical.

Pure truth and falsity, and by extension the entire contents of the objectivist and subjectivist perspectives, are easily seen as mutually exclusive. However, 'glutiness' and 'gappiness' have something in common in their presentation of a 'third view' that does not align with normal experience. Niels Bohr is quoted for an "oft-repeated dictum" regarding contradiction at the deepest levels of knowledge, which seems relevant to this point: "The opposite of a correct statement is a false statement. But the opposite of a profound truth may well be another profound truth."[208] We also find a distinction in the Pyrrhonic philosophy between 'true and false impressions'—*phantasiai*—and 'incomprehensibleness'—*acatalepsy*. Again, this is mirrored in the Buddhist doctrine of two truths, which distinguishes the conventional or relative truths, which relate to the realm of duality, from the ultimate or absolute truth of *nonduality*.

The dual levels of truth, at this point presented conceptually, rather than ontologically, can be found within the dialectical matrix, and can be understood by making a distinction between 'essence' and 'form'. These terms have historically been conflated by Aristotle's definition of substance, but I am using them more literally than Aristotle; I am also using them heuristically, so the terms are somewhat arbitrary. Here, 'essence' is the essential quality, character, or spirit of a thing—that which evokes its sense of meaning, and that which makes it intrinsically what it is. 'Form', on the other hand, is the shape, appearance, or structure of a thing—how it is rationalised or interpreted, and how it is realised in the world.

To provide an illustration in terms of normative action, the essence of a good deed might be to help someone, *or* it might be to make oneself look good. The form of the action may be the same, while its essence is different. On the other hand, different actions may be done for the same intention. However, my meaning operates at a level deeper than this, and is related to the psychology of *bias* in our theories. The presence or absence of bias in a theory determines its essence, while how that bias is realised determines its form. Unbiased perspectives always have the same form, though we may be unbiased for different reasons, while a biased perspective always has the same essence, though our bias may have different outcomes. This distinction will become clearer as we move on.

When we observe the dialectical matrix, we see a dialectical opposition between objectivism and subjectivism, and also between abjectivism and superjectivism, but these opposites are opposite in different ways. Objectivism and subjectivism are structurally opposite, and therefore have opposing forms, but the essence of them both is the same. That is, both are predicated on discrimination and bias between the two aspects of our being, it is just that this bias is extended in two different ways. We normally think of positions like materialism and idealism, or anarcho-capitalism and state socialism, as direct opposites, and in a certain sense they are. However, in a deeper sense, they are based on the same motivation, which is to isolate a particular aspect of experience and maximise its expression in our conception of the world.

Conversely, abjectivism and superjectivism are structurally similar while having opposing essences. That is, neither expresses any bias towards either side of experience, but each validates this neutrality in dialectically opposing ways. For the abjectivist, the reality in either is not apparent enough to discriminate between them, and for the superjectivist, there is too much reality in both. This is reflective of the fact that abjectivism is grounded in the belief-disrupting properties 'mutability' and 'indeterminateness', while superjectivism is grounded in the belief-satisfying properties 'immutability' and 'determinateness'. This is also present in the distinction between pessimism and optimism, which is mirrored by Chantal Mouffe's distinction between antagonism and agonism in regard to social values. The *form* of social value, and particularly the plurality of values, can be agreed between adversaries, but the essence of social value is perceived in opposing ways, the former as a conflict requiring compromise, and the latter as a contrast requiring collaboration.

The superjectivist perspective also validates the complementary logical principles of excluded middle and explosion, resulting in triviality, where every statement is true. Abjectivism, on the other hand, validates the logical principles of non-contradiction and implosion, while also asserting *no* statements as true, and providing a solid ground for scepticism. Note that the Pyrrhonic sceptic has no trouble accepting the law of non-contradiction, but they do the law of excluded middle.

Paul Kabay, who is notable for *embracing* trivialism, has argued that trivialism shares with scepticism the Pyrrhonic ideal of *ataraxia,* which means 'peace' or 'equanimity' and is achieved through the suspension of judgement. "For the pyrrhonist it is a paucity of belief that will free us from anxiety. But for the trivialist it is an abundance of belief that will free us from anxiety."[209] 'Ataraxia' is another idea that Pyrrhonism shares in common with Buddhism, where 'upekṣā' also refers to a kind of equanimity in the face of changing appearances.

Kabay also considers trivialism in the context of a secular spiritualism, claiming that the acceptance of everything might facilitate a merging or dissolution of the self within the ultimate reality. If trivialism is true then our world is such that all possibilities are contained therein, and Kabay suggests this as "a non-theistic concept of deity". He refers to Nicholas of Cusa as an adherent of this kind of thinking, which is worth mentioning since it was Carl Jung's study of Cusa that helped inspire his employment of the coincidence of opposites. Cusa writes:

Because the absolutely Maximum is absolutely and actually

> all things which *can* be (and is so free of all opposition that the Minimum coincides with it), it is beyond both all affirmation and all negation. And it is not, as well as is, all that which is conceived to be; and it is, as well as is not, all that which is conceived not to be. But it is a given thing in such way that it is all things; and it is all things in such way that it is no thing; and it is maximally a given thing in such way that it is it minimally.[210]

Jung himself expressed the same idea in his pseudonymously titled *Seven Sermons to the Dead*—a Gnostic text penned during his 'confrontation with the unconscious' period. In it, he discusses the nature of the *pleroma*, or 'fullness', which is the primordial reality in Gnostic and Neoplatonic cosmology. He writes:

> Nothingness is the same as fullness. In infinity full is no better than empty. Nothingness is both empty and full. As well might ye say anything else of nothingness, as for instance, white is it, or black, or again, it is not, or it is. A thing that is infinite and eternal hath no qualities, since it hath all qualities.[211]

If the supposed sceptic wishes to *assert* that all statements are not true, rather than *reject* that any statements are true, then under classical logic they are no different from the trivialist. This is because a glut and a gap are equivalent via De Morgan duality and double negation equivalence, and it is also part of the reason why Jc Beall argues for truth predicate pluralism. There is also a conceptual connection between 'everything' and 'nothing' for in each all contrast is dissolved, which is to say, they share the same *form*.

Consider that I draw some ink onto a page, and as I draw, an image becomes revealed amongst the ink. The image becomes clearer as more detail is applied, but there will come a point where applying more ink will serve only to obscure, and as the final spaces of the page are filled, suddenly it becomes blank once again. However, it is also *not* blank; it is both minimally *and* maximally filled. Consider also that we have a matrix of dots that is infinitely dense. Either the matrix encodes all possible patterns, or it encodes no patterns at all, just as a blank slate. To perceive pattern within the fullness would require an act of creation just as much as it would one of discovery.

Total presence and total absence are perceptually and conceptually equivalent, for it is only in difference that our perception has a content, and it is only in difference that our conception has a meaning. Without difference there is no 'that which is' apart from 'that which is not', only the 'that which is and is not', which is also neither. In other words, to perceive *everything,* as the union of all positive and negative space, without any contrast between them, is to perceive *nothing,* for perception is empty in the absence of form. Similarly, if a word or concept meant all things, then equally it would mean nothing at all. Nevertheless, there *is* a difference between them, but it is a difference in *essence,* as the information encoded within them is empty and indistinct.

This idea can be found wherever exclusivity and hierarchy form a part of the definition of the notion in question. For example, something is special just in case not everything is special, for if everything was special then nothing would be special. Likewise, if everyone is a winner then no one is a winner; if everyone is rich and famous then no one is rich and famous; and if I believe every statement, then equally, and for the purposes of rational discrimination, I believe no statement at all. Both Hegel and Heraclitus made use of this quirk. The idea that both everything and nothing is special can be interpreted only as a ubiquitous uniqueness, which is close to Hegel's interpretation of the concrete universal as an identity-in-difference, or a unity-amid-diversity. There is also Heraclitus' proclamation that the only constant in life is change, or *flux*, which can be identified with *his* conception of the concrete universal—the *Logos*.

Abjectivism and superjectivism can be viewed as two approaches to the unity of opposites, and their similarity arises in all domains of philosophy. We have discussed it in regard to logic and political theory, and we are moving towards an image of the ontological interdependence of these dialectically opposed characterisations of Being. As a final point of note, Stephen Palmquist has elucidated the connection in regard to epistemology and moral philosophy.

Palmquist observes that Kant's practical reflection, which serves to uncover the moral imperatives, and particularly the implications of the categorical imperative, yields synthetic a priori knowledge in Kant's system. However, when considered from a practical standpoint, rather than a theoretical one, Kant "must be allowing these terms to take on significantly new meanings". 'Synthetic' becomes 'a posteriori' for it is dependent on "the instantiation of practical ideas *in experience*", and 'a priori' becomes 'analytic' because it is being used to define an *idea* or *concept* of reason in which experience is contained. Palmquist thus concludes that:

> Whenever Kant says something like 'X is synthetic a priori, though only from a practical standpoint', we can interpret this as meaning 'X is analytic a posteriori'...Accordingly, the most accurate statement of Kant's position is that, whereas speculative reflection attempts to establish the synthetic a priori status of metaphysical knowledge-claims, hypothetical/practical reflection admits that the epistemological status of such claims cannot (and need not) be anything other than analytic a posteriori.[212]

Both the abjective synthetic a priori and the superjective analytic a posteriori involve recognition of the relation between the subject and object, but while the former is concerned with the necessary conditions *for* experience, the latter is concerned with the nature of experience and existence itself.[213] Again, we find this sentiment expressed by the dialectical matrix, for while superjectivism is predicated on the notion of self-reference and self-consciousness, abjectivism is predicated on the nature of things in the absence of bifurcation into the opposites of conscious experience.

From Nothing to Something

The nature of the oppositional relation between abjectivism and superjectivism is itself in opposition to the oppositional relation between subjectivism and objectivism. The entire dialectical matrix is self-referential in the sense that it is a dialectic of dialectics, a duality of dualities. The higher-level opposition exists between essence and form, for the dialectic between abjectivism and superjectivism is a dialectic of essence, while the dialectic between subjectivism and objectivism is a dialectic of form. This is because, as we have discussed, the neutrality of the former makes them equivalent in form, while the biasedness of the latter makes them equivalent in essence. Each element of the dialectical matrix can therefore be interpreted in terms of its relations, and this provides a basis for the idea that the dialectical matrix is a universal structure, much like a schema, which applies to all relevant philosophical notions.

To illustrate this a little better, let us develop the dialectical matrix for the notion of 'mathematical sign'. Here, our dichotomies are 'negative' versus 'non-negative', and 'non-positive' versus 'positive'. As with all previous cases, bias is expressed horizontally, so to speak, between subjectivist and objectivist perspectives, and neutrality is expressed vertically, between abjectivist and superjectivist perspectives:

	Non-positive	Positive
Non-negative	Non-positive & Non-negative $\sqrt{-1}$	Positive & Non-negative $+1$
Negative	Non-positive & Negative -1	Positive & Negative $\sqrt{1}$

Bias is related to the formal dialectic, while neutrality is related to the essential dialectic, and this is expressed by the fact that negation reverses the sign in the former but not in the latter. That is, pure negativity is the negation of pure positivity, but 'neither-positive-nor-negative' and 'both-positive-and-negative' remain neutral under negation. Perception is tied to a distinction in form, and 'everything' and 'nothing' are perceptually indistinct *because* they have no distinction in form. In logic, distinction in form is expressed by the fact that different statements have different truth values, just as formlessness is expressed by all statements having the same truth value. The dialectical matrix thus expresses the relationship between the formed and the formless, the perceivable and the imperceivable, the something and the nothing.

Let us suppose for a moment that the dialectical matrix really does describe the basic structure of reality. If the nothing and the everything—the abjective and the superjective—are aspects of a single source, then their separation provides a possible dynamic for the emergence of form from the formless. That is, the separation of the subject and object, as the emergence of a distinction in form, is interdependent with the separation of the abject and superject, as the emergence of a distinction in essence.

Speaking in terms of the dialectical matrix for the notion of sign, the positive and the negative can be constructed by splitting and recombining the two aspects of the formless, which are the non-positive-non-negative and the positive-negative. The non-negative aspect of the abjective and the positive aspect of the superjective together give the non-negative–positive, which is the objective; and the non-positive aspect of the abjective and the negative aspect of the superjective together give the negative–non-positive, which is the subjective. It is in this way that we can get the something from the nothing, but also the everything from the something. This process was expressed in Chapter Fifteen, with our construction of the dialectical matrix for non-classical logics, as well as with our example of resolving the gap and glut between two intaglio and relief structures through a mutual exchange. As we shall see in Chapter Twenty-Two, the driving force for this process would be the development of self-knowledge, the rationalisation of the paradoxical, and the necessary teleological actualisation of the concrete universal.

If this explication of the dialectical matrix has any bearing on reality, then phenomenal experience, so as the rational, is a relativisation of the absolute, whereby the incomprehensible is split up into two domains that are themselves comprehensible. This process ensures that nothing is ever really gained or lost in experience—nothing is created nor destroyed—and the abjective and superjective are imminent in experience rather than transcendental. Change is an expression of reconfiguration, and the positivistic and objective domain of empirical reality and sense perception is continuously balanced by and correlated with the negativistic and subjective domain of logical necessity and ideas.

To reiterate, everything that is positive and determinate is positive only insofar as it has been severed from the negative and immutable. The opposites depend on each other, and every presence must be complemented by an absence. If we recall the dialectical matrix for the notion of truth, the subjectivist perspective is syntheoretic with the assertion of falsity, which expresses an absence within that which is *not* false, and so for the dialectical matrix to be a representation of reality, falsehoods and absences must be things that exist in much the same way as do truth and presence.

Negative Being

Typically, when we call something true, what we mean to say is that an assertion corresponds to something real in the world which *makes* that assertion true. In the case of falsehoods, we accept the negation of this—that there is *not* a truth-maker for a claim. However, to *assert* the negation, which is to say, to assert that a state of affairs *is not* the case, such as to assert that 'there is not an elephant in this room', there must be something in the world which makes this true. That is, the absence of an elephant must be a genuine fact, and so we have to explain in what sense absences are to be regarded as things that exist. Bertrand Russell, notably a positivist, saw no other way but to accept the existence of negative facts that we may talk meaningfully about the world, despite the tension it places on a positivistic outlook. He remarks: "There is implanted in the human breast an almost unquenchable desire to find some way of avoiding the admission that negative facts are as ultimate as those that are positive."[214]

The obvious target for the location of such negative facts is in the mind, implying that mental properties are not simply reducible to facts about the brain. Indeed, the dialectical matrix suggests this approach, for the falsity predicate, which is also the designated value of a falsificationist paraconsistent logic, assumes the same part of the dialectical matrix that gives rise to abstract universals, and the mind more generally. Note that, just like negative facts, universal facts are not made true by positive facts, since no number of particular positive facts, like 'this swan is white' or 'that swan is white', suffice to justify a universal claim, like '*all* swans are white'.

How then do we assert a universal fact or a fact about a mental quality? We do so through negation, which is the essence of Karl Popper's falsificationism. 'All swans are white' is true just in case no swans are black or blue or green or pink or brown, and so on. The dialectical matrix suggests that universals are generally defined by negation because they *are* quite simply absences. That is, the universal 'whiteness' is not some mystical property existing in a transcendent realm of ideal forms, but the absence of all those things which are *not* white. As such, when we conceptualise the whiteness of a particular thing, we are recognising everything else that is different from it, and we comprehend the quality only insofar as we have this recognition.

Whereas concrete particulars are localised to a limited time and space, and so their conception is that of addition to, or presence within, emptiness, abstract universals evoke the whole of existence and subtract from it. Ideas are defined in this way, having no strictly positive properties, but rather being holes within wholes. Since ideas are the referents of the terms of our language, words also gain their meaning via their relations with other words. That is, words have no meaning by themselves, just as negative facts have no corresponding objects. Words have meaning purely because they do not mean the same things as other words, and we could understand no word without understanding the difference between one word and another.

The Swiss philosopher Ferdinand de Saussure, now considered the father of modern linguistics, came to this same conclusion. In his 1916 *Course in General Linguistics*, Saussure explained that both words and concepts "are purely differential and defined not by their positive content but negatively by their relations with the other terms of their system". Together, the 'signifier' and the 'signified', which form the linguistic sign, do yield a positive content, but language as a whole finds meaning purely in the isomorphisms between distinctions in ideas and distinctions in symbols.[215]

The possibility of extracting meaning from things that do not exist implies that such absences have some kind of being—that the meaning *refers* to something, and that there is more than one sense in which a thing can be called 'real'. Typically, 'positive reality' would be taken to mean simply 'reality', while 'negative reality' is just 'nonreality'. However, the dialectical matrix implies that Reality, as a concrete universal, extends in both positive and negative directions, and is a union of positive and negative constituents. This would align with our previous definition of Truth as the union of truth and falsity, and with a plurality of status predicates that account for the distinctions between paracomplete and paraconsistent forms of acceptance and rejection respectively. In this vein, Alexius Meinong, an Austrian philosopher, asserted that things can *be* without existing—that there *are* non-existent objects—since all intentional acts must have a content. If we can think about a unicorn, then that thought must be *about* something, even though that something does not physically exist.[216]

With this, we are now able to make a first step towards the recognition of the dialectical matrix as an ontological reality, and with it the establishment of paraphilosophy as the resolution to the problems of philosophy. Fundamental philosophical notions, such as knowledge, being, value, right, and truth, can be represented and interpreted in two conflicting ways. Objectivistic knowledge comes from the verification of what *is*, and subjectivistic knowledge comes from the falsification of what *is not*. The relation between the 'is' and 'is not' reflects the relation between the 'neither is nor is not' and the 'is and is not', and all four form a uniform structure that permeates both our perception of reality and the possibilities of our conception of it. The structure encapsulates our theoretical intuitions, the logical relations between them, and is also an empirical outcome of the philosophical endeavour. In the following chapter, we shall see that the structure is also quite simply the self-containing, and self-validating, totality of Being itself.

Chapter Twenty

The Proofless Proof

Isomorphism & Structure

Insofar as syntheorology demarcates the basic structure of philosophical notions and our conceptions of them, it provides a powerful tool for formalising analogies between distinct areas of discourse. If diverse interests submit to the same structure, then an argument, implication, or discovery in one might have profound consequences for another, which it may not have had access to in the absence of syntheorology. *Isomorphism* is the equivalence relation that validates syntheorology as a study of structure, and thus the establishment of the dialectical matrix as that structure. In the eyes of syntheorology, philosophical topics are like different languages through which our capacity for reason is expressed. As we shall see in the present chapter, this is one way to provide a definition of 'paraphilosophy'—as the preconceptual object that syntheorology describes, and the actual existent behind the appearance of structure.

Isomorphism, as the structure of an effective analogy, enables the transfer of meaning between different systems, languages, and indeed philosophical perspectives. In mathematics and geometry, isomorphism has already revealed itself to be of great interest and importance, but we are yet to realise its import for philosophy and the possible science thereof. Gödel's theorems stand and fall with the power of analogy, for in revealing the isomorphism between formulas of formal systems and Gödel code numbers, consequences were derived that did not seem possible before.

Graham Priest maintains that the isomorphism between various instances of self-referential paradox, particularly the set-theoretic and semantic varieties like Russell's paradox and the liar, entails that they all share a 'uniform solution'. Moreover, since hierarchical solutions, such as Russellian type theories and Zermelian set theories, would be nonsensical if applied to the semantic paradoxes, the isomorphism suggests that the solution to the set-theoretic varieties must be non-classical. As we have discussed, Priest favours a paraconsistent approach and the acceptance of dialetheia.[217]

In both *Gödel, Escher, Bach* and his later *I Am a Strange Loop*, Hofstadter spends considerable care to elucidate the importance of formal analogies as permitting the transfer of meaning between levels of our cognition, such as between neurological networks and higher-level 'symbols'. As he notes in the latter work, such transfers of meaning are present between isomorphic systems whether we are interested, or it is useful for us, to see them or not:

> A remark made with the aim of talking about situation A can also implicitly apply to situation B, even if there was no intention of talking about B, and B was never mentioned at all. All it takes is that there be an easy analogy—an unforced mapping that reveals both situations to have essentially the same central structure or conceptual core.[218]

In the years following Gödel's theorem, a growing emphasis on structural relations, above intrinsic identity, led to the development of several new and fruitful approaches in the philosophy of mathematics, such as category theory and homotopy type theory. The thematic thread was that isomorphic structures express a genuine identity relation, whereby everything we can say about one version of a structure should be equally true when talking about another.

The French Bourbaki group of mathematicians asserted that isomorphic structures should be considered identical in a literal sense, expressing their relation with the equals sign. In homotopy type theory the univalence axiom asserts that "logical identity is equivalent to equivalence",[219] where equivalence is a generalisation of isomorphism; and category theory generally is a mathematical formalism of structure in terms of mappings that preserve the functions and roles of elements within different categories or sets. Also, American logician Steve Awodey defines 'the principle of structuralism' for mathematics as the assertion that "isomorphic objects are identical", where identity is an equality of properties. He presents the principle as "the sharpest notion of identity available" for mathematical purposes and leaves the question as to whether mathematical isomorphisms are *literally* identical open to interpretation.[220]

Structuralism in the philosophy of mathematics is one arm of a larger philosophical movement that emerged in France at the start of the 20th century. Like Saussure's structuralism of language, by which it was influenced, the structuralists asserted that most, if not all, systems of human thought, feeling, and action are governed by structural laws, that structures are real things, and that objects have no properties outside their positions in structures.

On first sight, there appears to be something in common between structuralism and syntheorology as a methodology, and I think there *is* a way in which they are similar, though first I should highlight a significant distinction between them. Structuralism is largely oriented around the definition of human activity and perception in terms of structures, and this is because it considers such activity to be unnatural and conditioned by those structures. Conversely, syntheorology has no preceptive commitment to structure, nor to the idea that structures in perception are resultant from structures in cognition. Moreover, structuralism seems to imply the existence of a kind of Kantian noumena, which is obscured by the appearance of structure, while syntheorology is strictly analytical and unconcerned with any inferences or conclusions to be made.

The Real, Psychoid & Abjective

The major similarity between structuralism, or post-structuralism, and syntheorology can be found in the structuralist reinterpretation of psychoanalysis provided by the French psychiatrist Jacques Lacan. Lacan proposed a three-fold relation within the psyche of man, composed of 'the symbolic', 'the imaginary', and 'the real'. These terms are loosely related to language, perception, and the prelinguistic, which in Saussurean terms would be the signifier, the signified, and the unsignifiable. The 'imaginary' forms the bridge between our internal and external worlds, relating to the inbound field of impressions and primal instincts and perceptions that are absent of a rational structure. It is in accordance with the 'symbolic', which is based on language, that the imaginary becomes structured into a meaningful and rational experience.

Standing opposed to the dialectic between the symbolic subject and imaginary object is what Lacan calls 'the real', which is the preconceptual and undifferentiated noumenon, or thing-in-itself, which forever resists symbolisation and realisation through structure or form. Being a homogeneous whole that transcends the oppositions of ordinary speech, Lacan's 'real' is ineffable, unthinkable, and empty. Like the logical expression of the abjectivist perspective, and particularly the sceptical truth value 'neither true nor false', the real precludes form. It is described both as total absence—an unconceptualised nothingness—and as total presence—"impossible, superabundant plenitude", recapitulating the indistinctness between everything and nothing.[221]

There is thus a possible connection to be drawn between the symbolic, imaginary, and real from Lacan, and the subjective, objective, and abjective of syntheorology. A particular affinity between the real and the abjective can be seen via Julia Kristeva's 'abjection', as that which is cast away from experience. More specifically, 'abjection' refers to a recognition of the uncanny or horrific in the sense of that which disturbs the order and familiarity of structured human experience. For the abjectivist perspective, it is an absence of a mental content, and manifested in experiences of existential emptiness, extreme nihilism, or ego-death, but is not necessarily horrific. For Lacan, the real is "the essential object which isn't an object any longer, but this something with which all words cease, and categories fail, the object of anxiety *par excellence*".[222]

For Lacan, the horror of the real derives not from a dissolution of being per se, but of one's own subjectivity. The real is that which lies beneath the object but nevertheless gives reality to it, and horror is bound up with the fact that the underlying reality is inaccessible to our concepts. In contrast to this, the syntheoretic abjective is not divorced from structure, but is rather an integral part of it, and so structure is not simply an artefact of the subject and the symbolic order. The abjective is not the absolute ground of the objective, and its location on the dialectical matrix means that it can neither be described as positive, true, and existing, for it is not concrete, nor as negative, false, and non-existing, for it is not universal. The abjective is not a transcendent or noumenal thing-in-itself but is rather found immanently in the relation between the subject and object, mediating their separation and ensuring their continual correlation.

The nature of the abjective can be spoken in terms of abstract particularity, which moves syntheorology away from Lacan's structural psychoanalysis, but towards Jung's analytic psychology. Jung's archetypes of the collective unconscious, which condition the contents of the *personal* unconscious (which were once conscious contents), are related to the structures of Lacan's symbolic order. Jung's equivalent to 'the real', and that which transcends the archetypal representations of the psyche, he calls '*the* archetype', being in a sense *the archetype of the archetypes*.

Like 'the real', 'the archetype' is pre-conscious and preconceptual, and it is also that which the physiological mechanisms of the brain dissolve into just as do our psychic contents. In other words, the archetype is neither mental nor material, but something neutral to them both.[223] Jung distinguishes the neutral nature of the archetype from the psychic and the physical in his labelling it as 'psychoid':

> The position of the archetype would be beyond the psychic sphere, analogous to the position of physiological instinct, which is immediately rooted in the stuff of the organism and, with its psychoid nature, forms the bridge to matter in general. In conceptions and instinctual perceptions, spirit and matter confront one another on the psychic plane. Matter and spirit both appear in the psychic realm as distinctive qualities of conscious contents. The ultimate nature of both is transcendental, that is, irrepresentable.[224]

Jung's position is a form of dual-aspect monism, with mind and matter resting on a more fundamental stratum of reality which is itself neutral. If we were to assume that objectivity and subjectivity—the positive and the negative—are extrapolated from the abjective and the superjective—the two 'essences' or aspects of the neutral—then syntheorology would also suggest something along these lines. However, syntheorology merely describes the relations between these perspectives, and so does not dictate that the abjective, or psychoid, should enjoy any exclusively fundamental ontic position.

On the other hand, syntheorology does present a symmetry and parallelism between the two aspects of form, as entailed by the fact that extracting form from neutrality requires extension in two complementary ways. This is comparable to the zero-energy universe hypothesis in physics, which conjectures that the quantity of positive energy in matter is precisely matched by the negative energy of gravity, such that the total energy of the universe is 'neutral'. Similarly, Jung posits that mind and matter flourish in symmetry out of the transcendental, first as the archetypes that underpin our ideas and the natural laws that govern matter, and later as the conscious contents of our subjective and objective experience.

In collaboration with the Austrian physicist and Nobel laureate Wolfgang Pauli, the pair attempted to penetrate to what they called the *unus mundus*, or 'one world', with an attack from both sides. As they saw it, the dissolution of the conscious mind into the depths of the collective unconscious is mirrored by the dissolution of the determinate object into the quantum mechanical field of potential. The complementarity of particle and wave in physics mirrors the complementarity between the conscious and unconscious mind, both distinguishing the known, observed and manifest from the unknown, unobserved and unmanifest, and thus hinting at a shared essence. Pauli explains:

> Physics and psychology reflect again for modern man the old contrast between the quantitative and the qualitative... On the one hand, the idea of complementarity in modern physics has demonstrated to us, in a new kind of synthesis, that the contradiction in the applications of old contrasting conceptions (such as particle and wave) is only apparent; on the other hand, the employability of old alchemical ideas in the psychology of *Jung* points to a deeper unity of psychical and physical occurrences...To us...the only acceptable point of view appears to be the one that recognises *both* sides of reality—the quantitative and qualitative, the physical and the psychical—as compatible with each other, and can embrace them simultaneously.[225]

Abstract Particularity

The union of the quantitative and qualitative aspects of our experience, which Pauli and Jung allude to, coincides with the elements of the abjective perspective within the dialectical matrix, particularly in the forms of abstract particularity and the synthetic a priori. As we have seen, the paradigmatic examples of abstract particulars are numbers, and numbers present to us in both quantitative and qualitative forms. As quantities, numbers facilitate our investigation and representation of nature, and our physics becomes more abstract and mathematical as we penetrate deeper into the structure of matter towards the quantum realm. Max Tegmark even takes this fact as evidence that the universe is ontologically mathematical at a fundamental level. As qualities, numbers encapsulate basic conceptual principles—wholeness, duality, creation, order—and there is a rich history leading back through symbolic geometry, mysticism, alchemy, and back to Pythagoras and Plato, investigating the qualitative significance of mathematical and geometrical forms and relations—relations we encounter daily in art, architecture, and music.

It is no surprise then that, in his later years, Jung too came to the tentative conclusion that the nature of the psychoid lies in number, and he was in fact planning a work on the archetypal significance of numbers before his death. "Number helps more than anything else to bring order into the chaos of appearance... It may well be the most primitive element of order in the human mind...Hence it is not such an audacious conclusion after all if we define number psychologically as an *archetype of order* which has become conscious."[226] Marie-Louise von Franz, a Jungian analyst who worked closely with Jung, elaborates:

> Jung tells us that when we take away all properties such as size, consistency, and colour from an outer or inner object, what remains as what we might call its most elementary property is its number. The primary means for ordering something in the chaotic multiplicity of appearance is therefore number...*It is the most basic element of order in the human mind.*[227]

She continues:

> The numbers are probably the most primordial archetypes that there are; they represent the actual matrix of the archetypes…As an archetype, number becomes not only a psychic factor, but more generally, a world-structuring factor. In other words, numbers point to a background of reality in which psyche and matter are no longer distinguishable. Jung was particularly impressed, for example, by the fact that the Fibonacci number series corresponded to laws for the growth of plants, as well as by what Eugene Wigner called "the unreasonable effectiveness of mathematics in the natural sciences".[228]

The idea that number is "the actual matrix of the archetypes" alludes to the structuring process of experience and conception that is expressed by our *dialectical* matrix. Even more interesting is the fact that number can be expressed by the matrix in a manner reflecting the notion of *sign,* discussed in the previous chapter. Here, positive and negative numbers relate to the objective and subjective; imaginary numbers—the square root of minus one and all that—relate to the abjective, being neither negative nor positive; and zero is related to the superjective, being the point on the real number line where negative and positive meet. Note that, considering that the superjective is related to the notion of fullness, the union of a positive and a negative of equal magnitude gives zero. We could also argue that all real numbers—positive and negative—sum to zero, while on another view, all real numbers may sum to infinity, either positive or negative. Considered as a concept rather than a quantity, 'infinity' is therefore another candidate for the superjective, for the division of a number as the divisor approaches zero gives a limit at infinity, and this infinity is both positive and negative since zero can be approached from both above and below.

The imaginary unit i is the solution to the equation $x^2=-1$. Here, x can be neither negative nor positive since both –1 and +1 square to +1; i is therefore perpendicular to the positive–negative spectrum. When imaginary numbers were discovered in the 16th century, they were regarded as meaningless fabrications and make believe in relation to the reals. They were also called 'impossible numbers', mirroring Lacan's labelling of 'the real', and it was not until they were shown to be effective in describing certain aspects of nature that they were accepted as perfectly normal elements of mathematics.

In an essay written in 1916, relatively early in Jung's career, he endeavours to explain the function of the psyche which is capable of transcending personal subjectivity to mediate an exchange and integration with the unconscious mind. He referred to the conscious exercise of this function as 'active imagination', and it is reminiscent of the abject encounter with the real in psychoanalysis. Jung compared this function to the complex functions in mathematics which mediate the relationship between real (conscious) and imaginary (unconscious) numbers.[229]

'Active imagination' informed much of Jung's work after his break from Freud, and it played a part in Pauli's thinking as well. Hundreds of visionary dreams were recorded by Pauli, and many were interpreted and published by Jung and von Franz. In a dream of particular note, Pauli is met by a female piano teacher who removes and presents to him a ring from her finger. It is the "ring *i*", she tells him. Pauli responds, "The makes the void and the unit into a couple," and, "The ring with the *i* is the unit beyond particle and wave, and at the same time the operation that generates either of these."[230] Make of this what you will, but keep in mind that Pauli's work on the spin of subatomic particles provided a usage for the imaginary unit in physics, and for which he was awarded a Nobel Prize on the recommendation of Einstein.

Self-Similarity

Any connection between imaginary numbers and the psychoid or abjective is speculative at best, but there have been several thinkers who have attempted to take the idea further. Clinical psychologist Terry Marks-Tarlow attests that recursive sequences of imaginary numbers express the fundamental structuring pattern of nature, and "the self-reflexive underpinnings of the dynamic unconscious."[231] Note that functions of complex numbers—numbers with a non-zero imaginary and real component—generate fractal geometry such as the Mandelbrot set—the Buddha-like fractal that is often regarded as the apotheosis of mathematical beauty:

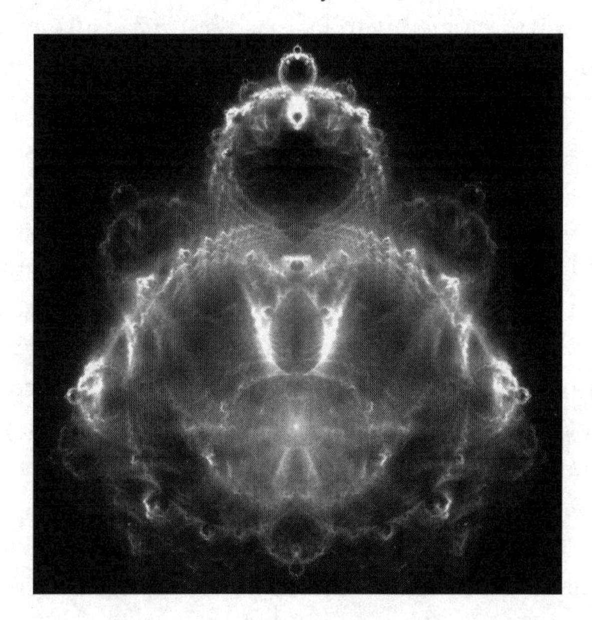

Marks-Tarlow provides evidence that fractal geometry is present at the lowest levels of both neurological and psychological functions, where it acts as "the bridge between mind and matter" in providing an intersubjective boundary of entanglement between conscious (real) and unconscious (imaginary) contents.[232] "Whereas Jung speaks of number as an archetype of order *which has become conscious,"* she speaks of "fractal geometry as deep order under chaos, *which is yet to become conscious and impossible ever fully to do so"*.[233]

We have already discussed the relationship between isomorphism and identity, and since fractals are formed of scale-invariant isomorphisms, Marks-Tarlow considers fractal geometry and self-similarity to represent the sign of identity in nature—that which maintains the structural equivalence of parts and wholes of a system. A similar point was central to Hofstadter's thesis in *Gödel, Escher, Bach,* where the subjective software level is simultaneously built upon and embedded within the objective hardware, producing a 'tangled hierarchy' that allows for scale-invariant changes encapsulated by the notion of the fractal.

Fractal geometry iterated from the interaction of the real and imaginary planes provides us with an explanation of the relation between the subjective and objective domains without having to posit the independent existence of a transcendental psychoid realm. We have seen that the abjective is syntheoretic with the absence of truth and falsity—existence and non-existence—and therefore can only express an insubstantial domain of possibility. It is through the presence of conscious experience that the abjective is divided and given both a positive and negative being, through the mutual relation of the subject and object. It is two in its extensions, but nevertheless remains whole in the boundary between them, where it plays the role of the dynamic structure of the *process* of evolution, rather than its substance. The development of the concrete universal ensures the fractality of the structure, for it exists both within and without it. That is, self-consciousness exists within the tangled relation of the subject and object, while also being that within which the relation is contained.

Much like the universal set in naive set theory, the self-participation of the concrete universal is a consequence of the fact that *all* things participate in the concrete universal. Understanding this universal as self-consciousness, it is necessary that all phenomena, whether abstract, particular, or both, are known only by their presence within self-consciousness. We thus have a situation where the superjective is an irreducible element of the matrix, which is conceptually dependent on its neighbours, yet also that which oversees and contains the matrix as a whole, along with itself within it. Following Hofstadter's illustration of consciousness as a tangled hierarchy between the objective brain and subjective mind, it is also simultaneously product and producer of the structure, and so the dialectical matrix reveals itself as a fractal hierarchy:

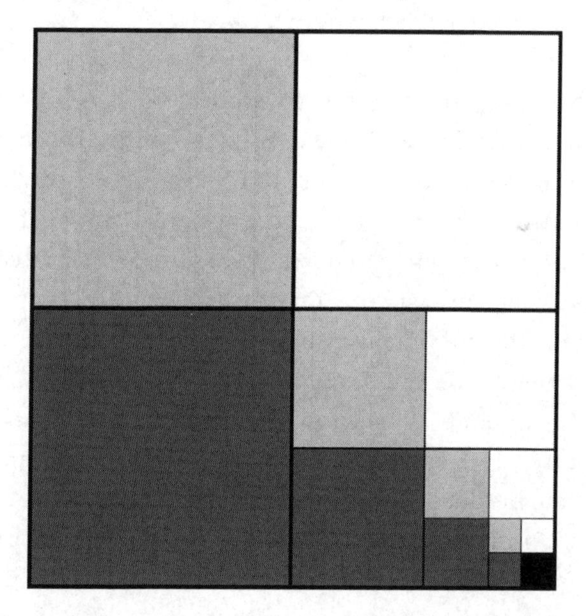

At one level, we can see that the superjective is an element of the structure, while at a higher-level, the structure itself is just the superjective—it is both within and without. We can conclude therefore that the dialectical matrix is precisely an image of self-consciousness, being the concrete universal. We can also infer this from the fact that the superjective is necessitated by the structure of the matrix, and the interaction between the objective and subjective, yet when the whole matrix is located within the superjective, as shown above, there is no superjective to be found anywhere within it.

The self-containing self can never be, and never not be, located, for it exists at all locations, and all locations exist within it. It is the most obvious and most obscure aspect of our experience, for it is everywhere and it is nowhere at once. It is an infinitesimal vanishing point which creates and is created by itself, and so an account of the Self in terms of the dialectical matrix can only be given by taking the system as a whole to be precisely the Self. Perhaps this is what Jung was referring to in *Seven Sermons to the Dead*, where he wrote:

> We are, however, the pleroma itself; for we are a part of the eternal and infinite. But we have no share thereof, as we are from the pleroma infinitely removed; not spiritually or temporally, but essentially, since we are distinguished from the pleroma in our essence as creatura, which is confined within time and space. Yet because we are parts of the pleroma, the pleroma is also in us. Even in the smallest point is the pleroma endless, eternal, and entire, since small and great are qualities which are contained in it. Only figuratively, therefore, do I speak of the created being as a part of the pleroma. Because, actually, the pleroma is nowhere divided, since it is nothingness. We are also the whole pleroma, because, figuratively, the pleroma is the smallest point in us and the boundless firmament about us.[234]

From a more materialistic perspective, Lacan also spoke of the sense of self—here related to *superjectivity* as opposed to phenomenological subjectivity—as an emptiness. It is the subject which has become its own object, a signifier with no signified. As Deleuze paraphrases Lacan, "It is always displaced in relation to itself," and it "is not to be found where one looks for it". He uses the analogy of an empty square, as in those sliding puzzle games you may have played as a child, being that within the structure which allows other symbols to be filled with meaning, but which has no meaning itself; as soon as the empty square is filled, it vanishes from its place.[235]

Deleuze elaborates that this "wholly paradoxical object" is always present in the structure, and since the object is empty, it is always precisely the same, and therefore the common element that unites corresponding structures. In being that element in relation to which "the variety of terms and the variation of differential relations are determined", the possibility of recognising structure entails the "positing something that is not recognisable", implying that the inconceivable is in fact the foundation of structure, and thus the conceivable.[236]

Interestingly, Lacan equated this 'something' with the imaginary unit *i*, being neither positive nor negative but nevertheless having a function. On the other hand, psychoanalyst Jacques-Alain Miller, Lacan's son-in-law, "borrows from Frege the position of a *zero*, defined as lacking its own identity, and which conditions the serial constitution of numbers". Recall that for Frege the number zero is the extension of the concept 'being non-self-identical'. Moreover, axiomatic set theory defines the natural numbers recursively from the empty set as a placeholder, and so there must be at least one object that exists necessarily, that object being the set with nothing in it.

The paradoxical fullness of emptiness, and the confusion between the *i* and the 0, as representing this union, is reflected in the dialectical matrix. If the superjective is the self-containing container that is also the matrix as a whole, then the entire structure is empty, for it is filled with the abjective, while also being full, for it is filled with the superjective. All expressions of the concrete universal are empty signifiers, for they do not predicate of anything beyond themselves, and because they are empty of meaning, all expressions of the concrete universal cannot be distinguished from each other. Knowledge, Being, Value—all are one. However, no phenomenon has meaning if not in relation to the concrete universal, and so the meaningful is constituted in its nature by the meaningless.

The Self-Validity of Paraphilosophy

The isomorphic expressions of the dialectical matrix are united as they recess into the flowering structure of self-similarity, which is both the activity and reality of the Self. To be real is to know oneself and to act in accordance with the good, and the relativity of our self-knowledge is the relativity of our being and virtue. The fractality of the dialectical matrix entails that the structure itself is known and justified via the analytic a posteriori, as the epistemological component of the superjective perspective. Self-validation is not only the most ideal foundation for the building of a science of philosophy, but the only possible foundation as such, for there is nothing outside of it from which it could be built.

Synthetic knowledge rests on the validity of the particular and mutable, and a priori knowledge rests on the validity of the abstract and indeterminate; but the analytic a posteriori, its object being itself, relies only on its own instantiation, so that by the very possibility of self-reference, the Self has become justified. If logical deduction gives us necessary truths, and empirical induction gives us contingent truths, and if these forms of knowledge are duals of each other, then for something to be established as both analytic and a posteriori then it must be bloody true indeed. The analytic a posteriori proclaims that '*This is* analytic a posteriori knowledge,' being tautological just in case it is instantiated. It shouts, 'HERE I AM', 'I EXIST', and could never be doubted unless it were already known.

With this, we can draw a conclusion that is of paramount importance to the present work: insofar as the dialectical matrix is identical to the Self and the superjective, the entire dialectical matrix is justified via the analytic a posteriori. That is to say, analytic a posteriori knowledge is just the dialectical matrix's awareness of itself, as it occurs through its own self-consciousness. It is what I am doing right now in writing this very sentence, and what *you* are doing in reading it.

We can now give a first definition of what is meant by the term 'paraphilosophy', and how it relates to syntheorology and the dialectical matrix. As stated numerously in this work, syntheorology is purely a method for uncovering the structure of our own capacity for ideation, theorisation, and belief. It is an objective science in that it seeks only to describe what already is, in conception, and makes no philosophical conclusions about what should be, in reality. The product of syntheorology is the dialectical matrix, with which it shares a similar relationship as the relation between the periodic table and chemistry, or the standard model and physics.

Syntheorology does not have an ontology, and it does not discriminate between the idea that the dialectical matrix is an artefact of neuropsychology, or of some transcendental aesthetic, or whether it reflects the actual structure of being. Conversely, paraphilosophy has no method and is a result of the dialectical matrix's own positing of itself. This act emerges through the possibility of analytic a posteriori knowledge, and its very being is the possibility of its own self-conception. It is thus like the Henkin sentence whose own self-position is its own self-proof. Paraphilosophy can only be known to exist insofar as it can posit itself, and it can only posit itself insofar as it can be known to exist. It is neither a theory nor an area of study; it is simply the field of possibility in which all theory and study is found. It is the empty infinitude of creativity within us, and *what we are basically*.

Truth, as the concrete universal, subsumes not only every particular truth, but also every particular falsehood. In this way, any rational conception or understanding of Truth, like that which we might achieve through philosophy, can only ever be a half-truth—the positive aspect of Truth. The negative aspect of Truth is found in analogical expressions of archetypal patterns and forms; in coherent relations; in motifs, morals, and myths; in words and concepts defined via negation; in feeling. All of these enjoy just as much 'reality' as positive being, for they are the reservoir of meaning that is required for any genuine appreciation of reason.

Deeper than this: pure, self-referential, concrete Truth consists in the absolute nonduality between the factual and fictitious, where there is no longer any distinction between the 'is' and 'is not', and about which we have nothing to say. We needn't say anything about Truth, however, for it is not a property *of* anything in particular; it is the concrete universal—pure self-referential Being—which is precisely the nature of the Self. It cannot be distinguished in thought, for it is also the subject of thought, which we are aware of through thought's very presence. It is the Law from which all possible knowledge derives—the 'I am'. It is the pillar of being that penetrates through the centre of all experience; it is the union of the obvious and the obscure.

Traditional philosophy has conceived of the notion of 'truth' as that which corresponds to, or coheres with, a description of reality that surpasses the work-in-progress of our finite and fallible minds. In the very act of searching, we conceive of truth as something outside us—as something we might attain, or as something in principle we cannot. We conceive that there is *a way* which reality *is like,* and our theories are formed in search of its likeness. But now we are faced with the realisation that what we are attempting to describe is purely our own ability to describe, which is evolving as a result of our action. We must acknowledge that reality *is* a work-in-progress, and we are active participants within it. Reality is simply our evolving capacity to envisage *what could be,* and all perception is an act of creation, just as much as it is of discovery. Nothing is true that is not known to be true, and there is nothing to learn that has not yet been learned.

Paraphilosophy is neither a dualism nor a transcendental monism. Moreover, '-isms' generally entail a belief in a theory, and paraphilosophy is not a theory, and so not something to be believed. Paraphilosophy reverses the emphasis from the objects of belief to the actual process involved when belief is taking place—to the capacity for belief, as a creative endeavour, and by extension the full scope of its possibilities. If we are to trust Aristotle when he says, "It is impossible for any one to believe the same thing to be and not be,"[237] considering that beliefs exclude each other, then it would be nonsensical to profess a belief in paraphilosophy. Conversely, we cannot *disbelieve* in paraphilosophy, for if we could then we would disbelieve in our ability to believe, which is also nonsensical.

The paraphilosopher is they who has reinterpreted the function of their belief; who understands that more is gained in the detached observance, rather than in the ascription, of belief; who has realised that the ability for belief is *the real thing* that itself is always looking for. Paraphilosophy turns the mirror on itself and swallows up its own possibility, rejoicing in the incontrovertible fact that itself exists. What you can know, you do not need to believe, so you should not believe paraphilosophy, for you know it.

An Admission

Even without considering that the superjective encompasses the abjective, objective, and subjective, the structure of the dialectical matrix can be deduced analytically through the relations of the resolutions of paradox. Moreover, the dialectical matrix is an empirical fact of our philosophic history, verified via the analysis of texts; and the syntheoretic correlations of the matrix are a priori intuitions. Indeed, the dialectical matrix may well be the only philosophical model that can be justified via any form of epistemological argument. It is only the analytic a posteriori, however, that can transform the structure into an ontology, and this requires going beyond both argument and belief. Paraphilosophy requires us to surrender to the fact that *it is knowledge already*, and in the absence of prior principles, this is fundamentally an act of faith. Not a faith to trust what you do not know, but a faith to stand by your ability *to know*.

As blasphemous as this seems to critical thinking and the analytic tradition, there is an instinct in me that paraphilosophy *should not* try to be justified—that justifying paraphilosophy can only ever obscure it from our view. To explain the basic in terms of the complex is to extrapolate and to construct a false meaning, whereas the recognition of paraphilosophy, in and for itself, must be personal. This does not mean that it should be accepted on a choice, or that it is sacrosanct, or that it is immune to criticism. To boot, paraphilosophy should be critiqued in excess; it means solely that it should not concern us whether paraphilosophy is accepted or not, and especially not on the basis of a rationalistic attitude.

To be quite sincere, I do not *mind* whether paraphilosophy is a good model or not, for *minding* is philosophy. Paraphilosophy, conversely, is *not minding*, and not limiting the creative capacity. What I *do* mind is the goodness it could beget, for it is this that motivates me to produce this work. I realise that these consequences depend on my own personal sense of paraphilosophy, and that this sense is affective and has become my temperament. I am cognisant and accepting of the fact that I trust the power of paraphilosophy not solely because of reason, and that I cannot see things any other way. Regardless of whether I have done a good job or not in making a case for paraphilosophy, the feeling that underlies my attempt at converting it to language is perfect to me because of what *I* see in it.

I have no assumption nor expectation that anyone should feel this feeling too, but I want to be transparent in stating that I give my faith to paraphilosophy because I feel its meaning, and I do not yet know whether this is simply my temperament, or whether others will feel it too. Nevertheless, for those who care to listen, the best argument I can give as to why paraphilosophy should be pursued *as if* it were true, is because I am in love with it, and I want to share this love with you. Indeed, paraphilosophy is simply the 'I', which is the concrete universal and the nature of the Self. It is the self-participating Love of Love, and so to love paraphilosophy is also to love oneself.

Chapter Twenty-One

Remarks on Paraphilosophy

Streams of Thought

As we endeavour to make some early remarks on the consequences of exalting the dialectical matrix to the structure of reality, let us be reminded of William James' perspective on the bifurcation of philosophical opinion, and what it tells us about the nature of our belief. For James, differences in belief are tied to differences in temperament, and there is an innate insincerity in the philosophic endeavour for our temperament has no proper place within the forum. Paraphilosophy escapes this criticism because it finds meaning precisely in the structural isomorphism between psychology and philosophy. Paraphilosophy also presents the proper context for the nature of philosophy and belief, whereby our temperament gains access to the forum without diminishing the integrity of our ideas.

The pretence of rationality arises in philosophy because sincerity conflicts with the standards we set ourselves for rigour. A purely rational belief flows from premise to conclusion and is promulgated in conscious discourse, while a purely sincere and basic belief flows from conclusion to premise and is relegated to the unconscious and the overridden. There is no reason why the two should not be in alignment, but any alignment is contingent, and if you truly believe that you are only rational, then you simply have not looked deep enough. We may present that a conclusion is derived from some premise, but we may merely be using the premise to our advantage, being happy to sacrifice it when a better option comes along. We present that our premise is stronger than our conclusion, yet it may be our conclusion that leads us to our premise. Internally, we cannot help but beg the question, and we cannot present this in the forum, so that we must be insincere if we are ever to be respected.

Our insincerity is made conscious whenever we err—when our conclusion *does not* follow from our premise. This should, in principle, discredit our position for we have already claimed that it *should* follow from our premise, but we seldom surrender our position for we cannot instantaneously reconfigure our minds. We may even accept the negation of our premise, for we do not hold our deepest beliefs for any *specific* reason, but because they cohere with all the other ideas that populate our minds. Our deepest beliefs exist as the context for our ideas and determine the extent of our capacity for them. If our temperament is an ocean, then our deepest beliefs are the rivers, and our ideas are the streams. These things are fused with our temperament, so that it becomes a *skill* to perceive reality a certain way. Reason shall not change what we are good at doing, which is believing *this* belief better than believing others.

While we see our reason laid out plainly in our books and essays, we must realise that beneath it all is chaos and disorder. We must recognise the power and intelligence of creativity in building a representation of the world that moves us—that makes it seem *as if* it could be real—and we must recognise that this power lies within us all, throughout all its many expressions. The fool can be perfectly satisfied in his belief, even if he seems 'not with it' to you. The mind *wants* a satisfactory understanding *for itself*, so that it seems that there is something other than understanding to understand, and so that we cannot all be right so long as we disagree.

For philosophy, disagreement reveals the limits to our reason, but for paraphilosophy it is an aspect of its power. For paraphilosophy, it is the *capacity* to reason, which is the truth, rather than the outcomes of its labours. Philosophy takes place *within* paraphilosophy, and paraphilosophy is the superposition of all particular philosophies. It is the matrix in which our reason moves, as it navigates a path of least resistance from our temperament to our belief. Every one of the many paths we can take is an idea, but all paths together is the Idea of the Idea.

Critical thinking is the engine of reality, just as analytic philosophy is the architect of paraphilosophy. The world that we experience is the sum of all ideas, and these ideas are relatively true insofar as they compose our capacity to reason what *could be* absolutely true yet isn't. Nevertheless, paraphilosophy, as the dynamic of self-knowledge *in time*, is not a mere trivialism, for the relative truth of a given theory is related to the degree in which it has been tunnelled out of the matrix of all possible theories—it is related to the size of the stream. As such, theories are not true in and of themselves, but are *realised* as true through the activity of the intellect, and their presence in our discourse.

We see patterns in philosophy because there are paths in the structure of Truth that are faster or more easily traversed than others, and the more a path is travelled, the more that path stands out from the matrix. This process only happens fractionally within the academy, for the realisation of our ability to conceive is tied to the evolution of the intellect biologically. Thus, reality is correlated with our theories because we create reality that way; every discovery is an act of creation, and every creation is an act of evolution.

Recall that James observed that philosophers and philosophies can be loosely divided into two groups, which he related to 'tough-mindedness' and 'tender-mindedness' respectively. Gödel too came to categorise our views in this way, which he expressed in a 1961 essay concerning his views on the future of mathematics and philosophy. He writes:

> The most fruitful principle for gaining an overall view of the possible world-views will be to divide them up...In this way we immediately obtain a division into two groups: scepticism, materialism and positivism stand on one side, spiritualism,

> idealism and theology on the other...Apriorism belongs in principle on the right and empiricism on the left side... One sees also that optimism belongs in principle toward the right and pessimism toward the left...Another example of a theory evidently on the right is that of an objective right and objective aesthetic values, whereas the interpretation of ethics and aesthetics on the basis of custom, upbringing, etc., belongs toward the left.[238]

Gödel recognised that the formal evolution of our conceptions extends outwards in these two directions, likening it to the intellectual development of a child. He envisioned that one day there would be a new science that would decipher this division, and in doing so reveal that "the truth lies in the middle or consists of a combination of the two conceptions". He supposed that this science would be a continuation of Kant's "central idea", by which he is referring to the structuring of phenomenal experience by the a priori intuitions of the mind, and from his own time suggests Edmund Husserl's phenomenology as the potential solution. Recall that Husserl's ontology consists of individual moments of conscious experience—of *abstract particulars*—and so there is a possibility of reconciling phenomenology with Jung's hypothesis of the psychoid nature of base reality, and tropes with mathematical objects.

Phenomenology is the study of the structures of conscious experience from the direct perspective of subjectivity, and syntheorology could itself be viewed as involving a particular kind of phenomenology, for it studies the phenomena of ideas, beliefs, and theories. However, Kant's idea suggests that human subjectivity is incapable of attaining to the true nature of reality, and thus that there is a schism between the noumenal and phenomenal. Paraphilosophy would stand this idea on its head, for the structures of experience are self-constructed as a means to realise the fundamental and are not there to obscure it from our view. If the noumenon can be said to exist at all, then it consists in potential, neither exists nor does not, and is no more fundamental than the phenomenal world.

Edmund Husserl was himself influenced in his ideas by William James' *Principles of Psychology*, and there was an effort at one stage in phenomenology to recast this work as a proto-phenomenological theory of intentionality.[239] Like Gödel, James felt that there was harmony laying hidden between the opposites of philosophy, and moreover that mystical experiences point towards this harmony. In *The Varieties of Religious Experience*, he endeavours to investigate the psychology of such experiences, and sought out methods, somewhat successfully, to achieve some insights of his own. He writes:

> Looking back on my own experiences, they all converge towards a kind of insight to which I cannot help ascribing some metaphysical significance. The keynote of it is invariably a reconciliation. It is as if the opposites of the world, whose contradictoriness and conflict make all our difficulties and troubles, were melted into unity...This is a dark saying, I know, when thus expressed in terms of common logic, but I cannot wholly escape from its authority. I feel as if it must mean something, something like what the hegelian philosophy means, if one could lay hold of it more clearly.[240]

It Takes Two

Paraphilosophy provides an explanation for the appearance of duality in the world, for duality is the only way to relativise the Absolute while retaining its wholeness. As we saw in Chapter Nineteen, to extract form from the nondual and formless is to distinguish the existent from the non-existent, which are the positive and negative aspects of being. Just as a wave has both a crest and a trough, all form has that which is revealed and that which is hidden. On one side, we have the positive object, and on the other, the negative subject. Each has form only insofar as the other completes them, with all that the object is not being a part of the subject, and all that the subject is not being a part of the object. Both sides are always equal, for they are a blending of the sceptically empty and the trivially full.

There is a single pattern of activity expressed in two opposing ways, and there is a necessary correspondence between these two expressions. Any change in one is mirrored by a change in the other, but no change in either is the cause of any change in the other, which is to say, they are mutually acausal. It may appear as though, for example, physical injury causes mental pain, or that choices cause events, but these things are not in fact distinct; they form two sides of the same coin, as the saying goes. Only experience is changing, and the subjective and the objective are interpretations of the same.

The parallelism between the mental and material provides us with an incredibly powerful tool for describing elements of the world that are otherwise out of reach for internal or external experience alone. Some aspects of experience are bound to be more comprehensible as determinate properties of physics and sensation, while others are more accessible as immutable properties of psychology and feeling. There may thus be aspects of both that can only be identified via their correspondence with their other. That is, we can understand physics through psychology, and psychology through physics, and this opens the doors to rich new fields of interdisciplinary and collaborative arts that will break down the walls of fragmentalism and isolationism that have been built by the spirit of our time. We shall find our resolutions in irony, for progress will be made only through a reconciliation with what we have tried so hard to conquer.

Only time may tell the nature of such new sciences, but there are guesses that we can make of them. For one, if there is an ideal equivalent for every material thing, then there must be a psychical counterpart to every field of science. We already have the fields of psychophysics, psychochemistry, and psychobiology, which explore the relations between physical and psychical events as they occur about the organism, but there may also be relations present in the objects of perception, and in systems not immediately affective of our physiological and psychological states. In the other direction, there should also be dualities within the elements of the unconscious, perhaps underscored by the quantitative and qualitative aspects of mathematics.

If this view is to be trusted, it would follow that there is an ideal correlate for subatomic particles, states of matter, chemical elements, molecular compounds, and everything up to the level of mind and body, and then the universe at large. Carl Jung emphasised that the historical discipline of alchemy was in fact an effort to harness the connection between chemical and psychoanalytic processes of transformation. We might also suppose that elements or chemicals, like lithium or lysergamides, and by extension neurotransmitters, which produce effects on human subjectivity, might provide insights into these basic connections. Paraphilosophy could even provide great insights for medical science, particularly for the treatment of chronic conditions with both affective and somatic components. Beyond this still, it may provide a path to enhancing experience in already healthful subjects.

'Psychophysical parallelism' was a prominent theory in the seventeenth century, with Baruch Spinoza, Nicolas Malebranche, and Gottfried Leibniz all espousing similar views. While monism is often passed as being more ontologically parsimonious than dualism, in regard to the mind–body problem, I find psychophysical parallelism to be the most parsimonious view of all, for it evaporates the profound problem of mind–body interactionism. Until now, however, the idea has not achieved widespread appeal for we have had no satisfying explanation to account for its modality. Nevertheless, on the grounds of dual-aspect monism, Jung and Pauli professed a specific effect of such parallelism, expressed by their concept of *synchronicity*.

Synchronicity is not simply the correlation itself but depends on the *perception* of a correlation between subjective and objective occurrences. These perceptions are acausal, so that any physical experience is not the cause of a psychological correlate. The sense of meaning betrayed by synchronicities suggested to Jung that they were not mere coincidences, but rather pointed towards a common ground state referred to as the psychoid *unus mundus*. Synchronicity is, therefore, bound to be viewed as an ordinary manifestation of psychophysical parallelism to one who already believes in the unitary nature of mind and matter, or as a fallacious exaltation of statistically predictable phenomena by those who don't. To Jung's credit, he always did his best to avoid making claims in the absence of empirical data, and this is why he discusses synchronicity primarily as a psychological phenomenon. It is thus most apparent and meaningful when occurring in a clinical setting, and therapists still utilise related techniques to their sanative advantage today.

Pauli's approach was somewhat more metaphysical, and Thomas Filk suggests that Pauli may have taken the acausal correlations of quantum entanglement to be the physical analogue of synchronicity, and a possible "model for the relationship between mind and matter in the framework of the dual-aspect monism he proposed together with Jung".[241] In this framework, the wave view of matter would be complementary to the unconscious mind, and the two aspects are united in the psychoid or abjective realm whose nature is unmanifest potential. Subjective conscious reflection serves as that psychological process which actualises—makes conscious—the potentialities of unconscious contents, and objective measurement is the corresponding physical process which actualises—makes material—the potentialities of unobserved quanta.[242] Synchronicity and entanglement would then be artefacts of a prior unobserved unity.

Complementary Extremism

As we move on to consider some of the theoretical implications of paraphilosophy, we must observe that paraphilosophy dismantles a certain conservative methodological and ideological ethic that otherwise holds significant weight. We can refer to such a temperament as 'philosophical moderatism', as it is emphasised in particular by the abjectivist perspective. Philosophical moderates recognise the equal weight of both subjectivist and objectivist arguments and seek to find a reconciliation between them. In Chapter Six, we characterised our belief-satisfaction in terms of the properties 'immutability' and 'determinateness', and so the moderate is someone who is unwilling to put all of their eggs in one of these baskets.

A thoroughgoing objectivist, according to this idea, would be one who wishes to eliminate the mind, as the subjectivist would eliminate matter. In the dialectical matrix, these extremes are found at the outer edges, and as we move closer towards the abjective, our perspective becomes more accepting of its opposite. For instance, eliminative ontologies would be the furthest out, a little closer would be reductivisms, then forms of property dualism, then emergent dualisms, and then certain kinds of panpsychism would be the most moderate while still retaining a bias in one direction.

I refer to moderatism as a kind of ethic because it confers empathy, open-mindedness, and a willingness to entertain diverse opinions. Moreover, in some areas of philosophy, such as axiology and political theory, moderation can beget genuine moral outcomes. For instance, anarcho-capitalism involves the neglection of equality just as state-socialism does of liberty, and the push to moderation is a very real moral factor in civil life. Extremism in all forms is commonly seen as a vice and not a virtue, yet paraphilosophy responds that this is only the case under the presumption of monoletheism.

We shall explore this idea more as we move forward, but for now at least understand that paraphilosophy confers a complete schism between our internal and external worlds, and this schism must be carried over to our discourse. *Even if* the ground state of reality were ontically neutral and abjective, we do not experience it as such, and so neutral and abjective approaches to axiology and political theory do not apply to the very real appearance of duality. As such, there is no reason for moderation for there is nothing to be moderate towards. Mind does not need to be involved in matter, nor matter to be involved in mind, and our reason is not limited in formulating theories by acknowledgement of what they exclude.

Paraphilosophy reveals the underlying unity between moderation and extremity, and it does so by means of a virtual dualism, or a *complementary extremism*. Monoletheic moderatism can only ever be reductive—it can only take away from the objects of perception, by surrendering the immutability and determinateness that is found in conscious experience. Despite this, moderatism brings monoletheism as close as it can get to the poly-perspectivism that is required for truly dispassionate judgements, but it can only do so by intelligently formulating a pragmatic system of sacrifices. Accordingly, paraphilosophy presents us with an unusual truth: the more moderate the idea, the less true it is, and the more extreme, the truer. At the same time, the more accommodating the idea, the greater genius it conveys, while the more rigid and controversial, the lesser genius it needs. These statements hold true when considered in regard to monoletheic philosophy, from which we can derive the ironic paradox that the strongest ideas are the 'worst', and the weakest are the 'best'.

Abjectivism is an *extreme* moderatism in that, if both sides cannot succeed, then both shall fail. It is nevertheless an extremely thoughtful perspective, reflecting its syntheoretic correlation with Jung's 'introverted thinking', for in the absence of any direct realism of our perceptions it must construct for itself a new and unfamiliar image of reality. Pure reason and experiment have no access to the transcendental, and so we find ourselves only with intuitions articulated by the synthetic a priori. Abjectivism lies at the end of the road for any inquiry that surpasses appearances and penetrates into the unmanifest. We find this in the successors to Kant's general idea, in phenomenology, in theoretical physics, in structuralism, and to a certain extent in Jung's analytic psychology.

It is quite evident that Western thought tends to lean towards the object, while the East leans more towards the subject, but all efforts for conciliation lead away from extremity, for partisan consensus is ultimately untenable. The abjective is the inevitable attractor of the intellect, for not only do appearances dissolve into the abjective when we look deep enough within them, and not only is it begot by increasing openness and tolerance, nor either that the greatest genius emerges in its favour, but we are *actively forced* towards the centre by the presence of conflict, and our conflict is universal. This is seen particularly clearly in the fact that most of the world's politics evolve towards authoritarian individualism, under the banner of 'liberal democracy'. This is not an undesirable outcome, of course, for moderatism is always better than discriminatory extremism.

The abjective is, however, a trap for the intellect, and it is a deadly one for its effectiveness increases on the smarter of minds. The entrapment arises when we forsake the experiential in favour of what is not, for in doing so we are blind to the inclusion of the structure which creates the abjective within the concrete universal. If abjective potential is regarded as the ground of the actual, then the superjective is its sky, and the two project inwards to form our relative world of being, while at the same time being created by it.

A Framework for Philosophy

With syntheorology as our guide, we can now provide an initial description of the nature of our *worlds*. To clarify this, here is the dialectical matrix for the notions of temperament, logic, truth, knowledge, being, value, and right:

	Indeterminateness	Determinateness
Mutability	**Abjective** Introverted Thinking Paraconsistent Paracomplete Non-true & Non-false Synthetic A Priori Abstract Particular Objective Relative Authoritarian Individual	**Objective** Extraverted Thinking Consistent Paracomplete True & Non-false Synthetic A Posteriori Concrete Particular Subjective Relative Libertarian Individual
Immutability	**Subjective** Introverted Feeling Paraconsistent Complete Non-true & False Analytic A Priori Abstract Universal Objective Absolute Authoritarian Collective	**Superjective** Extraverted Feeling Paradoxical True & False Analytic A Posteriori Concrete Universal Subjective Absolute Libertarian Collective

According to this model, paraphilosophy, as the actualisation of the dialectical matrix, declares the objective world to be grasped by the synthetic a posteriori, to be constructed of concrete particulars, to have value relative to the subject, to legitimise liberty and inequality, and to be consistent but incomplete. From the perspective of the objective world, empiricism, materialism, moral subjectivism, and anarcho-capitalism are true and their antitheses are not. Standing opposed to, but in complement of, the object, the subjective world is grasped by the analytic a priori, is constructed of abstract universals, has absolute value, legitimises equality and authority, and is complete but inconsistent. For the subjective, analytic rationalism, idealism, moral absolutism, and state-socialism are true and their antitheses not.

As we can see from this system, the duality between positive and negative being is tied to the distinction between the fragmented and the whole, for while objective constructs are quantitative, autonomous, and defined individually, subjective constructs are qualitative, interdependent, and defined collectively. The objective perspective speaks the indisputable truth, establishing consensus reality and motivating scientism, but it cannot dictate what is false, and this is reflected in the failure of sceptical positivism to win the battle over the felt quality of experience. The subjective perspective, on the other hand, has a harder time making definitive assertions, but it knows what isn't false, and it gains its power from an ability to illuminate structure on the backbone of negation.

The objects of the subjective—concepts, qualities, and ideal forms—are negative existents, which is to say they *are* without existing physically. Within the context of words and signs, these objects adhere to Saussurean linguistics, having no positive properties and defined in terms of each other. Concepts are more than this, however, for they complement something material. They are involved in material things, but negatively so, such that the union of corresponding psychical and physical forms brings us back to neutrality.

A concept is the superposition of everything its instances are not, making them very large entities indeed. That is, to produce the positive form of an apple from an infinite field of neutral potential, what is left out by this extraction is everything that apple is not, which has now become negative. The sharpness of a knife is correlated with the superposition of everything that is not sharp and is not a knife. The redness of a rose with everything that is not red and not a rose. This is the distinction between intaglio and relief, or the figure and the ground.

Objective and particular knowledge is formed of truth, while subjective and universal knowledge is formed of falsity, solving the problem as to what are the truth-makers of falsehoods and universals. This distinction also dictates the basis of ethics, cleaving a division between the positive and negative good. The objective world is syntheoretic with moral subjectivism, so that the moral worth of physical actions is relative to subjective conditions, such as those that are determinable via empirical methods. Normative moral subjectivism dictates what is right, and thus what actions one *may* undertake ethically, but it has no claim to the wrong, and no power to forbid any action. On the other hand, the subjective world is syntheoretic with moral absolutism, and since objective value is itself a universal idea, the moral worth of ideas is a logical matter of fact insofar as their meanings are contained within this universal. As with all universals, the not wrong is defined in terms of the wrong, and so normative moral absolutism does not dictate what *should* be thought, but rather what *should not*.

It is on this ground that we can use syntheorology to uncover the structure of the ideal political system for any point in time, and I have confidence that this will prove to be an incredibly important aspect of paraphilosophy. Any practicable political theory of course requires extensive deliberation, and I am not a political theorist. I present the following purely in regard to what is suggested by paraphilosophy, as I interpret it, and give little thought to how it could ever be implemented. I hope I can do enough to get across the basic character of the idea and implore theorists to improve upon it as they see fit.

Bidoctrinalism

Back in Chapter Ten, we made a brief pass over the moderatism that arises as a natural forcing of consensus in politics. Theoretically, such moderatism is aptly characterised by Isaiah Berlin's liberal conservatism, due to its being grounded in the syntheoretic moral theory of value pluralism. The complementarity in political theory exists between liberty and equality, which are syntheoretic with moral subjectivism and absolutism respectively. The competition between these values has meant throughout the history of civil life that both are limited to retain stability. Berlin's position, but also that of less liberal forms of authoritarian individualism, are distinctly pragmatic in nature, being evoked not only by the accommodative temperament of moderatism, but also as the mean of wide-ranging differences of opinion.

The incommensurability of liberty and equality fuels the disparity of public opinion, and a government assumes the intractable task of finding balance between the two. The only outcome of this strife is authoritarian individualism, coupled with a normative moral pluralism, and this must be seen for precisely what it is—a reluctant compromise that will fail to satisfy anyone who has seen their bias to its end. Nevertheless, paraphilosophy reveals that conservatism is an understandable and unavoidable implication of monoletheic thinking, and that social and economic life expresses the same acausal duality that is present between matter and mind. By ruling this duality with a single authority, we can never realise our intentions, for to reduce opposites to a compromise is to diminish their essential nature.

In accordance with the schism between the ontologically positive world of matter, and the negative world of mind, paraphilosophy dictates a schism between systems seeking liberty and equality respectively. To paraphrase Einstein's comment on the dual nature of matter: sometimes we must use the one system and sometimes the other; we have two contradictory pictures of civil life; separately neither of them fully permits for the realisation of right, but together they do. The libertarian image applies to the concrete, for liberty is a determinable state of social life, and the subjectivity of value allows liberty to reach its extreme while equality is ignored. The collectivist image applies to the abstract, for equality is an immutable state found evenly in poverty or plenty, and the absoluteness of value allows equality to reach its extreme while liberty is ignored. These two images need not contradict each other when they operate in entirely disjointed domains.

To clarify, each of our worlds demands its own *extreme* political system: anarcho-capitalism for the material, state-socialism for the ideal. Individualistic and authoritarian interests are then defined by the dialectical matrix in terms of the syntheoretic notion of value. Material goods, like new cars and foreign holidays, are luxuries that have value relative to the subject, and are satisfied through libertarian individualism. Ideal goods, on the other hand, like wealth and well-being, are necessities whose value is objective and absolute, and are satisfied through authoritarian collectivism. I have tended to refer to this pair as the *facetious* and *serious* domains, but, of course, the phenomenal world is a tangled mess, and not all things are clear-cut between them. Bidoctrinalism should not be seen as a static system, but rather as a means of maintaining harmony with the state of our social evolution. As we shall see, this should mean that bidoctrinalism gradually materialises in the world before dissolving again some time in our future.

This is a radical idea, but not one that is entirely unfamiliar; conservatism and liberal politocracies already attempt to differentiate between the wants and needs of society through different kinds of policy. Bidoctrinalism merely takes this one step further by recognising that we can only get so far while the serious and facetious are governed by a single system of value. This value system is underscored by the concept of money and is regulated by the rule of law; applying it both to our wants and needs is the perfect recipe for injustice and greed. Bidoctrinalism is, in this way, an *idealised* conservatism, for the aptitude of a conservative politic tends towards bidoctrinalism as its ability to demarcate between the serious and facetious improves.

The major challenge with implementing a system of this

nature is that it requires a complete re-evaluation of our store of value. Wealth generated in the facetious domain could not be used to gain advantage over others in the serious, and so bidoctrinalism implies the complete dualisation of monetary powers. Most likely, the system would entail the use of two distinct currencies, with limited means of exchange between them. 'Serious money' can be used within the serious domain, and 'facetious money' can be used within the facetious, but no amount of riches gained from facetious success can afford you higher standards of healthcare or education and so on in the serious.

In the serious domain, work may be performed with renumeration in credit rather than currency, and serious goods are effectively traded in exchange for service. This service would be a public service, so that all members of society contribute towards the collective. Of course, legislating that all individuals must fulfil a humanitarian or philanthropic duty is very authoritarian, but there are other ways to move slowly towards such a system. For example, there could be a way to put facetious money to use in paying others to fulfil your own liability, inculcating a new type of industry in the process. We are also reminded that only a small percentage of the workforce is in the public sector, and as this percentage increases, there will be significantly less work to be done per capita.

Many careers in the public sector of course require education, training, and full-time attention, so there may also be a means of facetious renumeration for any work that exceeds one's duty. In fact, the workforce need not change significantly from how it exists today, and the current system of taxation already exists to reallocate currency from the private to public sector. The important factor is the bifurcation of our monetary instruments, and there are many ways to implement an interaction between them.

The result of a dualisation of capital would be that the

serious could remain as a solid store of wealth, where there is no inflation, no profit, and total public expenditure. The value that we currently place on money, which is the cause of many of our issues, would be dismantled, for our comfort would not depend on it, and in regard to serious goods there would be no such thing as the rich and the poor. On the other hand, the facetious domain would be unrestrained by policy, allowing for the total realisation of liberty within it. Its currency would be liquid, granting the free exchange of goods, and there would be no taxation on profits that are gained. There would also be no law, in the facetious, for it is not the domain of objective morality. Instead, the system would be self-regulated in an open competitive market, and in accordance with the principles of anarcho-capitalism. There may be incentives, but there are no punishments, and the only limitation is that the facetious does not overstep its bounds and impact on the serious.

Participation in the facetious domain is up to the individual and has no influence on one's basic standard of living. If we choose to partake in the game, then success can bring wealth, fame, and glory, but we need not fear for our lives if we should fail, nor are riches the only marker of success. Success in the facetious domain can never place one person above another, for when we all go back home, we shall lie in the same beds. Power is given a new meaning under bidoctrinalism, but this does not make it any less rewarding. On the contrary, facetious success can only be a greater pleasure, for it is clear from the start that riches and happiness are not the same thing, so that we shall never expect that they should come together. Passions shall be pursued for their own sake, without restraint or fears of regret, purely for the purpose of inspiring individuality, and invigorating individuation.

In summary, bidoctrinalism is an alternative to the forced compromise between liberty and equality that is necessitated by any moderate doctrine. Just as eliminative materialism and idealism are true of the objective and subjective worlds, pure liberty and pure equality are the authors of each domain of civil life. The conflict between charity and competition will always give rise to disparity if they are left to interact, for the socioeconomic value of grace and greed respectively are conflicted. They pull the rope of society in opposing directions, creating great tension in the system, and the way to relieve this tension is to cut the rope in two.

The only way to create a libertarian collectivist system capable of uniting liberty and equality is if everybody was an ideal moral agent, which, until a lot changes in our world, remains an impossible task. To boot, paraphilosophy dictates that liberty and equality will only be united in the same system when the objective and the subjective are united in perception, representing the complete actualisation of self-consciousness in the superjective. In the absence of a ubiquity of ideal moral action, bidoctrinalism presents the only way to instantiate these values in their extremes, without eliminating social difference or monopolising economic power.

I must reiterate, however, that I do not present this thesis on economic or sociological grounds, nor am I qualified to attempt to do so. I also do not present it because I *believe* it to be the solution to the troubles of our time, nor even that such a system is possible. I present it solely because it, or something like it, appears to be implied by the science developed in this work. There will be many problems to be found in the idea, some I see already, and some I surely don't. But there are also solutions to these problems—solutions that work, and solutions that do not. I ask that you do not take the incompleteness or implausibility of the idea as a reason to reject it outright, for it may one day become clear that it is precisely the kind of radical system we shall need. Indeed, it is precisely the kind of system that paraphilosophy demands *we do need*.

The way I see politics and the development of civil life can be boiled down to this: *social evolution is a two-player game*. We have the ideal of developing from a state of dependence and disparity to one of liberty and equality—from authoritarian individualism to libertarian collectivism—but there is no single path between them. It's like the dialectical matrix contains a hidden dimension—a pyramid rather than a two-dimensional square—and we cannot traverse the mountain between the opposites. The path to our left leads to liberty, and the path to our right leads to equality. Even if we reach a state where one of these values finds its extreme, we could never get beyond it to a reconciliation with its complement, because *evolution is a two-player game*.

There is a gate barring disparity from equality, and there is a gate barring dependence from liberty. The key to the former is found at the latter, and the key to the latter found at the former. These keys are like pressure plates, and if both keys are activated at precisely the same time, then both gates shall open, but there is no way to pass the gate on your own. Those with which we compete we must cooperate, so that they shall be not our enemies but our allies all along. There is no physical evolution that is not also mental; there is nothing discovered that is not also invented; there is no freedom that is not also determined; and there is no perfection of the capital that is not also a perfection of the social. Everything is already in harmony, and there is only the fight against it.

Chapter Twenty-Two

Phenomenology of Self

The Becoming of Individuality

The concrete universal, as the fundamental existent, is the dialectical matrix as a whole, and a fractal entity existing both within and without itself. By inquiring into the nature of the concrete universal, we can understand the process by which form, and the distinction between the subject and object, emerges from the formless to inquire into itself, just like we are doing right now. We understand the superjective to express the union between classical dualities, such as truth and falsity, matter and mind. We also understand the concrete universal to be the phenomenon of self-reference. As such, a phenomenology of Self would be a study of how the phenomenon of self-reference appears to itself through its own self-reference; that is, *meta-phenomenology*. Finally, the concrete universal is formless insofar as it contains no distinction between the real and unreal, and there can be no final representation of itself within itself, which would require such a distinction, and also some termination to its own self-reference.

There are many ways that we can characterise the concrete universal, and this may lead to the assumption that there are lesser concrete universals within the concrete universal which is the Absolute. This would be a mistake; the concrete universal is a property that posits its own existence, and things are called concrete universals insofar as they express this property. There is no characterisation of the concrete universal that is above any other, and reality does not emanate from any one in particular. The whole of creation is in harmony with itself, and each fragment is identical to the whole, much like a hologram. The phenomenal world thus emerges from the inside out, and from each individual 'I', rather than from some transcendent source. Indeed, there is ultimately no true distinction between the Absolute and the relative.

Self-consciousness is not therefore some disembodied spirit existing outside of our experience, but rather has its being purely through the experience of individuals. The Self is also not, therefore, immediately self-conscious, despite its nature being to be self-conscious. Similarly, and following Hegel's understanding of the concrete universal as Individuality, it is not immediately individual for there is nothing other than it. That is, an individual is individual only insofar as it differs from another of the same nature. This is the paradoxical nature of the concrete universal, for any attribute it might have, also it does not, for such attributes imply differentiations that have not taken place at the beginning of time.

The concrete universal *becomes* what it is in time, and since *it is* what *it does*, it is the becoming of becoming, and the unity of being and non-being. For self-referential Individuality, considered as Absolute, to be what it is, it must become differentiated. This is the modus operandi of Carl Jung's developmental psychology, for which the principle of individuation is the process of becoming an individual differentiated from others, and the essence of every created being. It is part of the nature of Being to become differentiated as the realisation of Self, and we, as the individuals of Individuality, and through our ambition to divide the world by our concepts, are reflections of this primordial drive.

Furthermore, for Individuality to be fully realised as a concrete whole, it must become substantiated by instances of its own nature, for the only individuality in sameness is the sameness of the individuality *of individuals*. In other words, the difference *within* Individuality, being we as individuals, is the same *in our difference from each other*. This sameness *is* concrete Individuality, for it is a sameness-in-difference, a unity-amid-diversity, which *expresses* sameness just in case *it is expressed by* difference. This is the core of Georg Hegel's philosophy, for which it is the process of an inner dialectic that is the essence of Being, and which sublates the opposites, returning the Self to unity. The fundamental nature of reality is a paradox whose undoing is evolution, for the formless Being is absolute, and evolution is relative. Just as there is no sense to the concrete universal without particularity, there is no sense to the Absolute without the relative.

The formlessness of the superjective means that complete self-knowledge is indistinguishable from total ignorance, and since knowing and being are one, the being of the Self is indistinct from non-being. The process of learning of and recognising oneself as this paradox is also that of self-creation, which is, as Hofstadter suggests, "Quite like the Henkin sentence, which, by merely asserting its own provability, actually becomes provable."[243] Also recall Fichte's assertion that "the self *posits itself*, and by virtue of this mere self-assertion it *exists*".[244]

The evolution of self-consciousness is a cycle that never falls, rather fading away to give way to the new, much like the Shepard Tone which continuously rises without getting any higher. At the beginning and end of time, the Self neither knows nor does not know, and both knows and does not know, its own nature. The resolution to the paradox lies in the perception of time, which separates the states of pure ignorance and knowledge, like a circle whose beginning and end are at once separated and connected. Evolution is an organisation of form atop the completely formed, which is no longer perceptible, having become formless, and making way for form anew.

The pinnacle of conscious representation of Self lies at the middle of this process, for it is there that individuals are most individual, and that the Self's creation has the most power to reflect on its creator as separate from itself. Self-knowledge continues to rise from this point, but as form becomes less distinct, it becomes a deeper wisdom of the unity between knowing and not knowing. The Self thus surrenders the will to know as it reaches its completion, fading into pure ignorance again, which is precisely perfect knowing. The origin of life is thus a great forgetting that depends upon its own future remembering, like electricity that will only leave its source when the circuit is closed. The Self is *always* learning, but while the first stage consists in the learning of its own existence, the second stage consists in the learning that the nature of its own existence lies in its not knowing its own existence, for it is thus that it may continue to create, and that it may continue *to be creation*.

Evolution of the Object

The relative truth of things in regard to a given space and time is a fragment of the concrete Truth, and there is not a certain *way* that reality must be. Rather, there are ways in which reality can present itself—paths it can take—in the evolvement of its own self-recognition. The mundane truths of life, by which I mean physical and psychological facts, are created and reinforced in concurrence with our ability to comprehend our internal and external environments, as well as the required preconditions that would allow our self-reflection.

The content of perception is the rationalistically constructed objective phenomena it finds itself within, and the context for perception is the meaningfully structured subjective phenomena it forms about itself, and each counterbalances the extension of the other. The former is represented in space as sense perception, the latter in time as intellectualisation. These are not, as Kant might say, pre-structured lenses through which we view the real world, but are spontaneous, resonant, self-constructions of the world itself, which forge a path between self-ignorance and self-knowledge.

The understanding of consciousness is always correlated

with the extent of what there is to understand, as its subjective and objective environments. The beginning of time is a state of pure ignorance and absence of form, and the first 'movement' into form is the beginning of the real, such that the objectively real begins along with its subject, which is life. There is only a world *out there* when there is something *in here* to perceive it, so that the millions of years 'before' life began were not physical happenings but structures of potential. Our rationalisation of this past is an extrapolation of what reality would be like had something been around to perceive it, based on what reality *is like* when there is.

The evolution of the realistic is then the separation of the objectively real and subjectively unreal from the abjectively neither real nor unreal by the superjectively real and unreal. There is no mystery to the origin of life for there is nothing before life from which life must organise. We can understand the development of reality through understanding the biology of early life and the intellect correlated with it. For instance, if the first perception was a subtle distinction between here and there, or doing and not doing, then the first thing to exist is a direction or a movement in space. When the perception of brighter and darker comes into being, light becomes real, and with the perception of louder and quieter, sound. Everything is the awareness of an idea and its representation in space and time.

The development of reality is like any other learning process; first arise the simple things and later comes complexity. What exists at any point in time is precisely what the Self needs to substantiate the context of its current state of understanding. This understanding has a rational aspect, which is the foundation for the deterministic evolvement of physical reality in accordance with the laws of physics, and it has a meaningful aspect, which is the foundation for the immutable structures of subjectivity in accordance with the archetypes of the psyche.

The Self is innately creative, and our history is a creation whose purpose is to justify the present state of understanding, but it is not the cause of the present, nor is the future its effect. Causality and law are artefacts of the perfect consistency of our power to create, and the present only evolves in accordance with perceived law because the perception of natural law is a necessary component of the development of our understanding.

Art and science are united in paraphilosophy, for the world is a fiction that has made itself real. Our past makes sense in the context of physics just as any great work of high fantasy provides a context that is not only conceivable and believable, but which also provides crucial meaning to the events taking place. Nothing is superfluous, and nothing is missing. Physics is an exercise in world-building with no holes in its plot, and the discovery of new physics is an expansion of our capacity for conception, for creation and discovery are one. As such, physical evolution is the objective expression of an abjective or neutral process and must be complemented by its subjectivist counterpart. This applies not only to the objects of perception, but also to the objective mechanism of biological growth, which we understand as the natural selection of genetic variations.

Neo-Darwinism may well be a complete theory in content, but it is incomplete in context. A subjectivist interpretation of evolution need not alter the results or predictions of the theory, but it will provide a perspective that will further elucidate the physical processes involved within it. The notion of teleology—being directed towards some end—in biology is commonly seen to be at odds with natural selection, but on this view any 'attraction' towards some goal or purpose must confer the same processes as Darwinian evolution and be indistinguishable from it. Finalism and mechanism must be complementary perspectives on the same processes.

Evolution is teleological because 'being' is prerequisite for 'becoming', and if there is nothing that we fully *are*, then there is nothing for us to *become*. Nevertheless, the predetermined end of evolution does not intervene in the natural process beyond the fact that its existence is why the process must take place. There is only one beginning, and there is only one end, and without them there can be no becoming at all. Does this mean that selection is guided? Not necessarily, it means only that we live in the particular actualisation of potential in which evolution *does reach* this end by chance alone. From our perspective, within the process itself, the fact that evolution does reach this goal when it could just as easily have arrived at some other life, can only be interpreted as miraculous luck. The same can also be said for all the other required conditions for life on our planet. This does not mean that anything has actively guided it, for this end is within the possibility of random events and needs nothing other than chance.

The correspondence between mechanistic and finalistic causes of evolution would be a natural consequence of the general epistemological principle that Niels Bohr envisioned for biology and other sciences. He saw that biology would require two independent modes of inquiry, one to describe the physical mechanisms of cellular processes, and one to describe the goal-oriented functions of the organism at large.[245] Pauli was greatly inspired by Bohr's ideas and though he remained conservative with making comments that confronted scientific reductionism, he did divulge some doubts regarding the traditional understanding of Darwinian evolution. In a letter to Bohr of 1955, he writes:

> In discussions with biologists I met large difficulties when they apply the concept of 'natural selection' in a rather wide field, without being able to estimate the probability of the occurrence in an empirically given time of just those events, which have been important for the biological evolution. Treating the empirical time scale of the evolution theoretically

> as infinity they have then an easy game, apparently to avoid the concept of purposiveness. While they pretend to stay in this way completely 'scientific' and 'rational,' they become actually very irrational, particularly because they use the word 'chance', not any longer combined with estimations of a mathematically defined probability, in its application to very rare single events more or less synonymous with the old word 'miracle'.[246]

Statistically improbable events become inevitabilities when considered in an infinite timeframe, and 'miracles' become predictions. In Pauli's view, agency and purpose are complementary to each other, yet chance occurrence is only relevant to the former. He therefore considered the existence of additional "laws of nature which consists in corrections to chance fluctuations due to meaningful or purposeful coincidences of causally unconnected events".[247] We should be able to operate from the perspective of mechanism, however, without losing essentially any information. Likely, this requires an alteration of the notion of 'chance occurrence' to one that contains an element of mechanistically inexplicable *luck*, yet which is nevertheless meaningful when considered from a complementary perspective.

There could not be any more an ideal factor for unifying the two perspectives of our world than that of *randomness*, for even if a selection is completely determined from one perspective, it may be entirely unpredictable from another, given a correspondence between would be determined and random events. Given that, to the best of our knowledge, wavefunction collapse in quantum mechanics is a truly random event, chance may also be the factor uniting many antitheses in philosophy, such as the one that exists between free will and determinism. If this is indeed the case, we can better understand the true message of Pauli and Jung's concept of synchronicity as meaningful coincidence, and how it relates to the parallelism between mind and matter.

On this view, it is not that thought or intention or will *does* anything actively to affect physical reality, but that they are the subjectivist manifestations of an abjectivist influence on probability that occurs within the realm of potential itself. From our perspective, any correlation between what we think and what we do—perhaps even between what we collectively will and what happens about us—is a lucky coincidence whose regularity is also a lucky coincidence. On this view, life and everything within it is a natural miracle, for it exceeds mechanistic explanation without betraying the modality of mechanics. So long as an outcome is within the realm of possibilities that we have, it is always possible that we may roll a hundred sixes in a row, and even if the outcome were predetermined 'from above', *we* can only put it down to chance.[248]

Evolution of the Subject

In the beginning stages of evolution, duality is at a minimum, which is to say that there is less of a distinction between psychical and physical domains of experience. This is reflective of the lower level of intelligence and self-recognition, comparable to those lower life forms present in the world today, for whom there is also less of a distinction between the subject and object. This early stage of evolution is simply an expansion of form, by which the expression of an external representation and an internal intellectualisation of experience gains clarity and focus.

The development of the intellect is correlated with the ability to perceive the realism of the objective world, which further down the line would bolster our adherence to objectivist and materialist accounts of the nature of such form. At slightly earlier stages in the human evolution, when the objective world is quite literally less real, and our intellect less capable, it is natural that we would be less inclined to mechanistic beliefs, and this is reflected in the historical records.

It is not merely that we become better able at perceiving the regularity and predictability of our objective environment, however, for the development of the intellect also carries with it an increase in the *desire* for rational explanations for experience. That is because experience is an *action* of the Self, and actions we can be better or worse at performing. As we gain skill in the action of experience, we gain interest in the action of experience, and thus the vision of mastering the skill, at which point we would have become a fully realised being.

However, the increase in our reason is not equivalent to an increase in wisdom, for wisdom is a recognition of the nonduality between the subject and object, while intelligence and reason are tied to the ability to divide them. It is often assumed, and for good reason, that the birth of modern wisdom occurred with the ancient Greeks, first giving way to natural philosophy, and later to physical science. Yet, the separation of the subject and object carries with it an incompleteness, and thus a *need* to know, and a pretension that one knows when really one does not. True wisdom is the lack of a need to know, and a recognition that knowledge without ignorance misses Truth entirely. Let us not forget the core teaching of arguably the wisest philosopher of ancient Greece. In Plato's *Apology*, Socrates concludes his being wiser than other men from the fact that none of them know anything of genuine worth, yet, while others think they know when they do not, Socrates neither knows nor thinks that he knows.

Paradoxically, knowing that one knows nothing is itself a kind of knowledge, and since the ground state of reality is a complete absence of form, absolute knowledge is indistinguishable from total ignorance. This comingling of opposites is the fundamental nature of the Self, and the academic facetiousness of the ancients may be warranted by a deeper connection they had to Being, which eclipses the need to know. What we would regard as 'mythological ignorance' today may be representative of the past contraction of the dialectical matrix and duality, which provided a meaning and intellectual satisfaction that no longer cut it as time went on.

When one is aware of the perfectly obvious, which is the absolute fundamentality of the Self, no energy need be spent on promulgating this basic fact. The awareness is instinctual, and as the intellect supplants the instinct, the identity of the Self becomes forgotten, and mythologies become archaic fantasies which say nothing of the nature of the real. This forgetting is a natural and necessary component of evolution, for it allows the development of the intellect and eventually the remembering of the Self. The intellect cultivates our uncertainty, and uncertainty drives the development of new inquiries and new technologies. These things symbolise our departure from the Self, but they are also that which will facilitate our eventual reunion with it. Conceptual thinking, discrimination, and differentiation shall be the wounders that heal.

Language is inseparable from this process, for it cultivates our individuality and facilitates diversity in our temperaments and opinions. Language is the modality of differentiation in our conceptions of nature, but also the means by which we seek their reunion. As we move further along the process of the expansion of the dialectical matrix, approaching the present day, the need for survival and the perceptual clarity of the object that supports it is diminished at the hand of increasing technologies. Our technology not only increases our chances of survival, but also increases the ease of communication, expediting the rate of information transfer and acquisition, and transforming the character of language along the way.

The path to self-knowledge first leads away from oneness and into form, enrolling consciousness in a training camp for rational thought, which will eventually be capable of recognising the destination as back where it once started. The key difference between remembering and forgetting is that while you don't know what you're remembering until you've remembered it, you *do* know what you're forgetting until you've forgotten it, and so the former is an unconscious process while the latter is conscious. If the evolution of our awareness of essence and Self is like a Shepard Tone, then the evolution of our awareness of form and duality is like a sine wave. It rises until it reaches the peak of duality, wherein the development of the intellect and the biology is capable of enough self-reference to instigate a grand revelation of Self. Then, the dialectical matrix, with its unyielding love for itself and its yearning to be whole, begins the long process of contraction again until the end of time.

Considering that the expansion of duality is the means by which the intellect is developed, so that it might recognise the Self, the contraction of duality is expressed by a reframing of the intellect from the role of analysing and differentiating, to one of synthesising and integrating. We venture out into the darkness to reclaim a great relic and bring it back with us into the light. The matrix expands in ignorance, drawing life and reason into form as the desire for reunion intensifies, and then contracts in understanding, relinquishing life and reason as peace returns once more. This is a great realisation of life, and one which is present in every moment we enjoy as living beings. In accordance with our fractal nature, this great truth is encoded into the breath—our most basic act of complementarity and being, and both a conscious and unconscious act. We inhale by our energy and effort, taking in the air that is our life, raising our awareness of the present moment, and creating a pressure inside us. We exhale automatically to find equilibrium, quieting the mind, releasing tension, and returning to the ground state of our being. This is not just an analogy. *Everything* is written in the breath, if only one would observe it.

Chapter Twenty-Three

Beyond Belief

Rivers of Thought

Throughout the history of our discourse, evidenced are the many ways in which the intellect can conceive of the objects of truth. Seldom does one man agree absolutely with another, yet remaining stable between them is the idea that their goal is static and complete. We feel that our ability to conceive ventures towards a truth that exists outside of us, and we build an image of our world on the ground of common sense. We place value through our thoughts onto this image, and we imagine that the image has value in itself. Our philosophy is always going somewhere, for it is always the *outcome* of our reason that is the thing of value, and if we do not find for it an object, we feel that our ability *to value* shall be wasted.

Paraphilosophy asks us to reimagine the purpose of our philosophy. For paraphilosophy, the *object* of our inquiries is the field within which our creativity and intellect moves, and our conceptions are passages carved out of this potential. To place value on these channels is to deny value to oneself, and to all those other conceptions we can feel but prevent ourselves from doing so. Paraphilosophy does not suppose that our conception has a goal—it is not work to be done, but rather an act of playful exploration. Value remains within the self, for the purpose of philosophy is to discover what philosophy can be done, not to uncover some end that would eradicate the search. Paraphilosophy is the full power of our conception and is exposed *by* philosophy. It posits nothing other than the exploration itself, for *it is* the exploration *of* itself.

To exalt an idea to the status of a truth—to *believe* that it is true—is to forsake the full power of the intellect, and to present a part as if it were a whole. As our attention flows through the channels of the mind, the path grows larger, and it gains new tributaries that connect with other paths. We become *better* at *doing* our belief, so that the same idea is never equally true for two people. Our beliefs become self-reinforcing feedback loops, and the more we understand how our idea makes sense, the less capable we become at seeing the sense in others. If we are truly skilful at our belief, then we could be skilful in rationalising others, but we never get a chance to see our creativity through, for our temperament and our habits have already defined us.

We guide our consciousness down a previously constructed path, triggering the same neural connections with each new pass. We find what we think about out in the world, being more sensitive to the familiar, and further solidifying the path of our thought. We build our belief structures as though they fill an empty space, not realising them as cracks in the whole that prevent the flow of consciousness from flooding the entire surface of the mind. To perceive, think, and feel paraphilosophically is to allow consciousness to take any path, to relinquish belief, to realise the full power of our ability *to believe*, and to *conceive* more than we ever *could* believe. This is not to say that every idea we can hold reflects the phenomena we see outside of us, but that every idea we can hold forms the potential from which our world is built, and from which it could have been and *could be* built differently.

The fundamental nature of reality is a paradox, while our beliefs are by their nature monoletheic. While we can believe two separate things that defeat each other, if we are to trust Aristotle, "No one can believe the same thing can, at the same time, be and not be."[249] Insofar as Truth is a union of what is and what is not, we cannot hold Truth as a belief. Truth is not something that might be the recipient of belief, but that which lights up potential and provides a status to it. Truth is the power to see, and to imagine that something other than itself *has* itself independently; it is our perfect creativity and the indefiniteness which is its nature.

Creativity is *not*-knowing, *not*-believing, *not*-being, and the only path to Truth lies in the alignment with this principle without the striving to align. Love, goodness, virtue, beauty, peace; they all come from the suspension of judgement, and this suspension is acceptance. This does not mean that our reason shall be thwarted; on the contrary, our reason shall be free, and may retain the rigour and reliability we have given it through our search for true and justified beliefs. The difference is that it shall be targeted at its own capacity, and not some absolute truth. To boot, the only absolute truth is that there is no absolute truth, and this we cannot believe.

Truth is not something that one can learn, nor something that can be told; it is something that we are doing, and which is occurring through us, even in the moments of our darkest ignorance. The closest we come to a rationalisation of Truth is in the knowledge that we do not know. Thus, I hope you are beginning to see, the semblance of truthfulness crafted in this work is only meaningful relative to the monoletheic belief. This is a work that examines the rhetoric of others without using any rhetoric to discriminate between them; yet, it is a work of a more subtle rhetoric, lurking amongst the indifference, and with a very particular goal in mind. This goal must ultimately defeat itself, so that if I am to do my job well, you will not *believe* a word that I have said, for they are but words, and words are not the Truth. The goal was never to give you something; the goal was only to take away.

Losing the Search

There is a long history, in both the East and West, of the existence of a covert perennial wisdom. The mystery schools in the West; the sacred religions in the East; the spiritualisms and esotericisms of the modern developed world. All these traditions covet something secret—a sudden revelation or gnosis that would unveil the mysteries of life. 'Enlightenment' is a name we have given to such an event, meaning literally 'coming into the light'. I do not want to undermine these traditions, but on the view of paraphilosophy, it is precluded that there should be some *spiritual* truth to which we can become 'enlightened'. Any mystical knowledge must be of the psychological variety, and thus concerns form.

The curious thing is, we can *know* that 'enlightenment' does not exist, and only when we know this, and not when we believe it, do we see the trick. The problem with the concept is not that it is useless, for indeed it has a use; the problem is that the concept fuels the delusion that it has a meaning, and that one should therefore strive to discover this meaning. The concept *does not mean anything*; it is an empty signifier in the most literal sense. There is only belief and its absence, and any belief that one might discover something more than belief is a belief like everything else. The secret is that there is no secret, and the absence can be found only by not seeking in the first place.

Now, let us use this term 'enlightenment' heuristically, in full recognition of its emptiness. If Truth is the simultaneous presence and absence of all ideas, and alignment with Truth is a suspension of belief, then the only way to 'enlightenment' is to dissolve the notion of enlightenment from one's mind. Non-belief would be the substance of 'enlightenment', being neither an idea nor an event, but rather a state of mind. It is the *basic* state of mind, which is the source of creativity, and on top of which all beliefs are planted. The concept is itself a paradox, for it is the knowledge of its own non-existence. We are chained by our own minds, by our concepts, and the concept 'enlightenment' is the strongest chain of all. For theists, it is the concept of a 'God'; for physicists, it is the concept of a 'theory of everything'. It is any notion of an absolute, for the Absolute is Paradox, and only when we see the perfect sense in the statement 'A iff not-A' shall we see that there is no gnosis which is not agnosis.

It is in this way that the empty can become full, and for those who see the hoax, the concept of 'enlightenment' becomes a hilarious pun, and the most ironic of all the ironies. However, for those who do not, it becomes a nasty trick. While there are surely many motivations for the spiritual path, which are valuable in their own right, the pursuit of the highest wisdom of all is an itch impossible to be scratched. Moreover, it is one which could cost you a lifetime; perhaps, according to some traditions, several lifetimes. Those who sit in meditation searching will never find of what they seek, and if they are lucky, they will realise why they *can't* one day, and thus they shall know that they have found it. Of course, they didn't really find anything, for they had it all along, and they just couldn't see it.

Success shall not be caused by spiritual skill, and it is only the glorified deception that there is something to gain that prevents one from *not* seeking for so long. Granted, this can be painted in a more positive light. If we are to assume that the trick is intentional, then perhaps it is actually effective—that only by searching hard, and searching long, and then by giving up, does *not searching* become *not needing to search*, which is an important distinction. Moreover, the concrete universal is so ordinary—Knowledge so inescapable, Being so immediate, Value so extensive—that to realise what we've got we must often lose it first. Every moment is evidence of the impersonal fundamentality of Self; and searching for it through words and practice is like diving to look for air.

The impressiveness of the natural state is dependent on how different one thought before it, but there is nothing innately impressive about the basicness of being, and one should beware of anyone who pretends as though there is. No man or woman has ever gained enlightenment; no man or woman will ever gain enlightenment; the Self is already complete. There are no masters, no buddhas, no saints; there is nothing to be found;

there is nobody to find it. The robe does not make the man holy, and the service does not make him see. The only lesson is one of silence, and silence has no speaker.

Impassioned Indifference

There is nothing wrong with holding ideas as if they were true; so long as they serve a purpose and are discarded when they do not, there is no distinction between the play and Truth. Attachment to one's ideas is only a choice, and if one sees oneself as those ideas, to surrender them would be to surrender oneself, and an impossible choice to make. Philosophers may find themselves particularly in this bind, for their task has been set *in search of* the truth. Of course, there is no incompatibility between the suspension of belief and the discernment of those relative truths which compose the dialectical matrix. It is rather the monoletheic assumption that some part of the relative is really the Absolute, which is the corrupter of Truth.

The search for wisdom is a self-referential negation, and one which reinforces the sense of a need to find, and to fill the hole which is made by digging. Those who never start the search get by just fine, so long as they do not search because they think they have found, but those who search endlessly now can never stop. If you are one of those people, *it's okay to stop now...* The only wisdom is peace, and peace is *not searching*. Moreover, peace does not destroy the play, it allows the play to be what it is, without having to question it.

The Pyrrhonists attested to this ideal in ancient Greece; 'ataraxia' they called it, meaning 'imperturbability' or 'equanimity'. Like paraphilosophy, they saw the suspension of belief, or 'epoché', to be the route to ataraxia, though their emphasis was negative—to "neither deny or affirm anything".[250] Pyrrhonism is a scepticism, which is an abjectivist perspective in paraphilosophy complemented by the trivial position of *both* denying *and* affirming. That is, for paraphilosophy, the emptiness of creativity is identical to its fullness, and the absence of any truth or falsity is the absolute freedom of the creativity. One must move beyond the distinction between not-doing, which is to say, neither denying nor affirming, and doing—denying *and* affirming—for it is in this that we may realise the nonduality between the Absolute and relative, essence and form. Thus, we shall not be empty shells for Truth, but self-conscious actors selflessly flaunting individuality in both defiance and gratitude of the empty.

Paraphilosophy here has something in common with the so-called 'ninth view' of Nāgārjuna's catuṣkoṭi, which is largely equivalent to the dialectical matrix for the notion of truth. This is the view which is not one of the four corners or their negations—not 'this view', not 'that one', not 'this view and that one', not 'neither this view nor that one', but rather all at once and none at all. Keep in mind that paraphilosophy is properly about philosophical theories, temperaments, and the elements of the dialectical matrix, and so paraphilosophy can be defined as not objectivism, not subjectivism, not superjectivism, not abjectivism. Paraphilosophy is rather all at once and none at all, for the dialectical matrix, which could be anything, is only the way that it is because it is the best way it could be.

This perspective is a rather nonsensical thing to try and get across in language, but I promise that there is sense to be found within the nonsense. Let me attempt to convey it through an analogy. Imagine that you have two small children, and they are quarrelling over which of their favourite cartoon characters is the best. You are interested in their debate, because you like the cartoon too, so you pick a side and argue in their favour. Yet, you do not *care* which character is best, and you partake for reasons *other* than being right. While your children are genuinely concerned about the truth, you have become too wise for the question and have therefore risen above it.

Now let us up the ante, to matters of fundamental philosophy.

It is possible to have the same view as you have to those cartoons in regard to human nature. You are not abstaining—you are passionately involved, but you are not intellectually invested in a result. Just as you realise the characters in that cartoon as the modality through which a deeper narrative is conveyed, you can realise our efforts in philosophy as the modality through which the Self is inverting its attention inwards, and that all happenings are a part of this performance.

This is an ability we all have, and it is 'the thing' that wisdom seekers are searching for—to not care about caring by caring about not caring about caring. It is not apathy—there is no shortage of interest—it is only the suspension of a temperamental prejudice, so that the libido may find value in *all* available prejudices, and particularly in the being able to find value in all available prejudices, which is the essence of *self*-value. It is not a lack of personality, but an ability to play without attachment, and it is not something that can happen overnight, but something one must work at. The more one aligns with this eternal temperament, the freer one becomes in making conscious decisions about the kind of character that one would like to play. The informant of these decisions is then the concrete universal, the sameness-within-difference, and the love for one's fellow man and woman.

All individuals are actors, and all actors have a drive to be individual. Some people have been playing a character for a very long time, so that they continue to play even when they are alone. This does not show a weakness, but a strength of the individuation process. Breaking the spell is the only weakness, for a weakening of the play is what is needed for a return back to the Self. Yet one must be weary of the trap of playing the character of no-character, for this is like holding the view of no-view as a view. The wisdom of impassioned indifference is much like the concept of 'coolness', for one can neither be wise nor cool so long as one thinks that they are. For the so-called master to present themself as a blank slate, to dress in robes of gold, to speak without an ego, or to encourage others to prostration, is simply uncool—unless, of course, that's precisely the point.

Passionately indifferent or not, the end of evolution shall work out just the same, for we are interminably attracted towards the light, which is our own self-knowledge. Those who break free from the illusion merely check out early, and they have no reason to evolve, for they have become already. It is not a bad thing to indulge in the illusion, but we outgrow it, like a child outgrows their toys. The real trick is to learn to like them again, for then you will have seen that *the illusion is the real*, and that *everyone* has a part of the truth. I don't have everything figured out, but I'm content enough to not obsess any more. I'm ready to get lost in the play, and to erase any folly that my thoughts might be true. This work has become far too real for me, and it's the unreal I yearn for now.

Notes

1. Dietrich (2011).
2. Nagel (1989).
3. Graham (2007).
4. Aristotle (1924).
5. Chalmers & Bourget (2021).
6. Popper and Eccles (1985), p. 51.
7. Hume (2008a).
8. Sextus Empiricus (1935), p. 179.
9. Hume (2008b).
10. Hume (1985), Book I, Part IV.
11. Kant (2007), pp. 18-19, B16-B17.
12. Ibid., p. 288, B355.
13. Hegel (1892), p. 100, par.48.
14. Hegel (1991), p. 20.
15. Hegel (2010), p. 381, par.11.286.
16. Russell (1999), p. 746.
17. Žižek (2013), p. 8.
18. Hegel (1997), p. 233.
19. Maxwell (1865).
20. De Broglie (1923). De Broglie's thesis was confirmed by the Davisson-Germer experiment in 1927.
21. Einstein (1938).
22. Bohr (1928).
23. Bohr (1937).
24. Wheeler (1963).
25. See Allinson (1998).
26. Bohr (1928).
27. Mazzocchi (2010).
28. Bohr (1933).
29. Bohr (1985).
30. Bohr (1949).

31. James (1890).
32. For example, Holton (1970).
33. See Allinson (1998).
34. James (1909).
35. James (2000), pp. 8-9.
36. Ibid., p. 9.
37. Ibid., p. 11.
38. Jung (1985), par.537.
39. Jung (2016), p. 285.
40. Ibid., p. 282.
41. Ibid., p. 362.
42. Ibid., p. 284.
43. Ibid., p. 4.
44. Ibid., p. 333.
45. James (2000).
46. Jung (2016), pp. 317-328.
47. Ibid., pp. 335-338.
48. Ibid., pp. 351-357.
49. Ibid., pp. 357-361.
50. Ibid., pp. 328-333.
51. Kristeva (1982).
52. Ibid., p. 2.
53. Eppinger & Hess (1915).
54. Eysenck (2014).
55. Jung (1992).
56. Lester (1983) and Lester and Berry (1998).
57. Benjamin et al. (1996).
58. Lienhard (2017).
59. Gazzaniga, Bogen and Sperry (1965).
60. Brederoo et al. (2017).
61. Semrud-Clikeman, Fine and Zhu (2011).
62. Benziger (2000).
63. Barnes (2002).
64. Paraphrased by Socrates in Sedley (2007).

65. Also, Hume's distinction between relations of ideas and matters of fact, and Leibniz's distinction between conceptual and logical truths.
66. Descartes (1971).
67. Locke (1997).
68. Mill (1884).
69. Putnam (2010).
70. Kant (2004).
71. Kant (2007), pp. 83-84, B73.
72. For example, Parsons (2001) and Osherson et al. (1998).
73. Frege (1999).
74. This is the 'Negative Way' in Parsons (1986), p. 83.
75. Smit (2000).
76. Ibid., p. 236.
77. Ibid., p. 237.
78. Plato (2005), par.91a-96c.
79. Plato (2009), par.74-76.
80. See Burgess (1983).
81. Field (2016).
82. See Loux (2017).
83. Mulligan, Simons and Smith (1984).
84. Williams (1953).
85. Husserl (2006).
86. Smit (2000), p. 255.
87. Ibid., p. 254.
88. See Tegmark (2014).
89. Stern (2007), p. 130.
90. Ibid., p. 132.
91. Ibid., p. 133.
92. Note that moral non-cognitivists and nihilists do not believe that there are such things as moral truths, so they are anti-realists but not subjectivists.
93. This is the analogy of the sun from Plato (2012), par.507b-509c.

94. Moore (1903), p. 15, par.13.
95. Kant (1998), p. 31, par.4:421.
96. See Rauscher (2002).
97. Hume (1985), p. 654.
98. Wong (1995).
99. Berlin (1998).
100. Wong (2019).
101. Berlin (1998).
102. MacIntyre (2007), p. 181.
103. See Rabinowicz and Rønnow-Rasmussen (2003), p. 395.
104. Firth (1952).
105. West (1990), p. 1473.
106. Ibid., p. 1486.
107. Iyer et al. (2012).
108. Carney et al. (2008).
109. Berlin (2007).
110. Schwartz (1990).
111. Smith (1776).
112. See Nozick (1974).
113. Rothbard (1975).
114. Lenin (2021).
115. Marx (1966).
116. Hayek (2007).
117. Cherniss and Hardy (2004).
118. Chomsky and Segantini (2015).
119. Aristotle (1924), par.1005b.
120. Ibid., par.1011b.
121. Ibid., par.1003a.
122. Ibid., par.1011b.
123. Schopenhauer (1909), p. 280 [286].
124. Venn (1880).
125. Boole (1854).
126. Ibid., p. 27.
127. Ibid., p. 42.

128. Stone (1936), p. 37.
129. Kneale and Kneale (1962), p. 413.
130. Russell (1999), p. 591.
131. Russell (1900).
132. Boole (1854), p. 6.
133. Frege (1990), p. 5.
134. Frege (1980).
135. Frege (1951).
136. Frege (1980), par.20.
137. Ibid., par.74.
138. See Boolos (1997).
139. Russell (1990a).
140. Frege (2013), p. 253.
141. Heck (1997).
142. Gao (2011).
143. Russell (1990b).
144. Hofstadter (1999), p. 22.
145. Ibid.
146. Priest (1994).
147. Nelson and Grelling (1908).
148. For example, a definition of the natural numbers required an axiom of infinity to ensure the existence of an infinite number of individuals.
149. Hilbert (1902).
150. Hilbert (1970).
151. Gödel (1990).
152. The non-disprovability of the continuum hypothesis was shown in Gödel (1940), and its non-provability was shown in Cohen (1963).
153. Shanker (1987).
154. Berto (2009).
155. Fogelin (1968).
156. Berto (2009), part 3.
157. Priest (2006), p. 46.

158. Drucker (1991), p. 96.
159. Priest (1979).
160. See Ibid.
161. Priest (2006), pp. 247-260.
162. Wittgenstein (1978), par.III.59.
163. Wittgenstein (1989).
164. Carnielli and Rodrigues (2020).
165. Dummett (1995).
166. Brunner and Carnielli (2005).
167. Ibid., p. 163.
168. Douglas Hofstadter uses a similar analogy in Hofstadter (1999), pp. 64-75, albeit for slightly different purposes, in terms of the distinction between figure and ground in two-dimensional imagery.
169. Beziau (2016).
170. Hernández-Tello, Macías and Coniglio (2020).
171. Tarski (1956), p. 155.
172. See Kripke (1975), p. 690.
173. Beall, Glanzberg and Ripley (2018), pp. 2-3.
174. Ibid., pp. 44-45.
175. See von Neumann (1958).
176. Wang (2016), pp. 186-187.
177. Lucas (1961).
178. Putnam (2011).
179. Lucas (1961), p. 121.
180. Hofstadter (1999), p. 559.
181. Ibid., pp. 584-585.
182. Ibid., p. 533.
183. Ibid., p. 708.
184. Ibid., p. 709.
185. Fichte (1982), p. 97.
186. Palmquist (1993a).
187. Palmquist (1993b).
188. Stern (2007), p. 121.

189. Kant (1998), p. 31, par.4:421.
190. Rauscher (2002), p. 482.
191. Ibid., p. 483.
192. Palmquist (1993a).
193. Plato (1956), par.330.
194. Ellerman (1988).
195. Plato (1993), par.10a.
196. Mouffe (2000), p. 5.
197. Ibid., p. 10.
198. Ibid., p. 5.
199. Hegel (1991), p. 189, par.142.
200. Ibid., p. 197, par.155.
201. Mouffe (2000), pp. 102-103.
202. Jung (1968), par.259.
203. Da Costa (2006), p. 117.
204. Ibid., p. 105.
205. Beall and Restall (2006).
206. Beall (2013).
207. Priest (2010).
208. Heisenberg (1971), p. 102.
209. Kabay (2008), p. 102.
210. Cusanus (1985), p. 9, par.12.
211. Jung (1989), p. 379.
212. Palmquist (1993a).
213. Palmquist (1993b).
214. Russell (1919), p. 4.
215. De Saussure (1959), pp. 116-120.
216. See Reicher (2019).
217. See Priest (1994) and Priest (2000).
218. Hofstadter (2007), p. 150.
219. Awodey (2013), p. 10.
220. Ibid., p. 9.
221. Minsky (1996), p. 147.
222. Lacan (1991), p. 164.

223. Jung (1973), p. 277, par.417.
224. Ibid., p. 280, par.420.
225. Pauli (1994), p. 259.
226. Jung (1973), p. 583, par.870.
227. Von Franz (2001), p. 44.
228. Von Franz (2001), p. 216.
229. Jung (1973), p. 97, par.131.
230. Quoted in Rosen (2014).
231. Marks-Tarlow (2013), pp. 27-28.
232. Ibid., p. 40.
233. Ibid., p. 41.
234. Jung (1989), p. 379.
235. Deleuze (2004), p. 185.
236. Ibid., pp. 184-187.
237. Aristotle (1924), par.1005b.
238. Gödel (1995), p. 375.
239. Levine (2018).
240. James (1982), pp. 387-388.
241. Filk (2014).
242. Mansfield (1991).
243. Hofstadter (1999), p. 709.
244. Fichte (1982), p. 97.
245. See Mazzocchi (2010).
246. Atmanspacher and Primas (2006), pp. 27-28.
247. Ibid., p. 31.
248. Considering that random events are determined by a non-physical process, the situation becomes even more curious when we consider rolling a hundred statistically random results in a row. That is because, it is not just meaningful events that are synchronistic, but all events as such, and we are unable to detect them as so. If this is true, it would require a total revolution in science, and a total reinterpretation of what we are investigating through the scientific method. When I first learned something about

quantum mechanics, it seemed perfectly natural to me that paraphilosophy implied not that locality or realism are violated, as in most interpretations, but rather the free choice of the experimenter. This idea is now explored in so-called 'superdeterministic' theories. However, while superdeterminism is usually taken to imply a total absence of free will, for paraphilosophy, it is the very mechanism of free will, and the modality of the synchronicity between mental and physical events. This raises many further questions as to the precise origin of determination—questions I think are answerable—though this is not a topic for the present work.

249. Aristotle (1924), par.1005b.
250. Sextus Empiricus (1990).

Bibliography

Allinson, R.E. (1998). Complementarity as a Model for East-West Integrative Philosophy. *Journal of Chinese Philosophy*, 25(4), pp. 505-517.

Aristotle (1924). *Metaphysics*. Translated by W.D. Ross. Oxford; New York: Oxford University Press.

Atmanspacher, H. and Primas, H. (2006). Pauli's ideas on mind and matter in the context of contemporary science. *Journal of Consciousness Studies*, 13(3).

Awodey, S. (2013). Structuralism, Invariance, and Univalence. *Philosophia Mathematica*, 22(1).

Barnes, J. (2002). Democritus. In: *Early Greek Philosophy*. London: Penguin Classics.

Beall, Jc and Restall, G. (2006). *Logical Pluralism*. [online] consequently.org. Available at: https://consequently.org/writing/pluralism.

Beall, Jc (2013). Deflated Truth Pluralism. In: C.D. Wright and N.J.L.L. Pedersen, eds., *Truth and Pluralism: Current Debates*. Oxford; New York: Oxford University Press.

Beall, Jc, Glanzberg, M. and Ripley, D. (2018). *Formal Theories of Truth*. Oxford; New York: Oxford University Press.

Benjamin, J., Li, L., Patterson, C., Greenberg, B.D., Murphy, D.L. and Hamer, D.H. (1996). Population and familial association between the D4 dopamine receptor gene and measures of Novelty Seeking. *Nature Genetics*, 12(1), pp. 81-84.

Benziger, K. (2000). *Thriving in Mind: The Art & Science of Using Your Whole Brain*. K B A Pub.

Berlin, I. (1998). On Pluralism. *New York Review of Books*, [online] XLV(8). Available at: https://www.cs.utexas.edu/~vl/notes/berlin.html [Accessed 29 May 2020].

Berlin, I. (2007). Historical Inevitability. In: H. Hardy, ed., *Liberty: Incorporating Four Essays on Liberty*. Oxford; New

York: Oxford University Press.

Berto, F. (2009). The Gödel Paradox and Wittgenstein's Reasons. *Philosophia Mathematica*, 17(2), pp. 208-219.

Beziau, J.Y. (2016). Two Genuine 3-Valued Paraconsistent Logics. In: S. Akama, ed., *Towards Paraconsistent Engineering: 110*, pp. 35-47.

Bohr, N. (1928). The Quantum Postulate and the Recent Development of Atomic Theory. *Nature*, 121(3050), pp. 580-590.

Bohr, N. (1933). Light and Life*. *Nature*, 131(3309), pp. 457-459.

Bohr, N. (1949). Discussion with Einstein on Epistemological Problems in Atomic Physics. In: P.A. Schilpp, ed., *The Library of Living Philosophers, Volume 7. Albert Einstein: Philosopher-Scientist*. Open Court, pp. 201-241.

Bohr, N. (1985). The Quantum of Action and The Description of Nature. In: J. Kalckar, ed., *Niels Bohr Collected Works: Volume 6*. Amsterdam: North Holland, pp. 201-217.

Boole, G. (1854). *An Investigation of the Laws of Thought*. London: Walton & Maberly.

Boolos, G. (1997). Is Hume's Principle Analytic? In: R.G. Heck, ed., *Language, Thought, and Logic: Essays in Honour of Michael Dummett*. Oxford: Oxford University Press.

Brederoo, S.G., Nieuwenstein, M.R., Lorist, M.M. and Cornelissen, F.W. (2017). Hemispheric specialization for global and local processing: A direct comparison of linguistic and non-linguistic stimuli. *Brain and Cognition*, 119, pp. 10-16.

Brunner, A.B.M. and Carnielli, W.A. (2005). Anti-intuitionism and paraconsistency. *Journal of Applied Logic*, 3(1), pp. 161-184.

Burgess, J.P. (1983). Why I Am Not a Nominalist. *Notre Dame Journal of Formal Logic*, 24(1), pp. 93-105.

Carney, D.R., Jost, J.T., Gosling, S.D. and Potter, J. (2008). The Secret Lives of Liberals and Conservatives: Personality

Profiles, Interaction Styles, and the Things They Leave Behind. *Political Psychology*, 29(6), pp. 807-840.

Carnielli, W. and Rodrigues, A. (2020). On epistemic and ontological interpretations of intuitionistic and paraconsistent paradigms. *Logic Journal of the IGPL*, 29(4), pp. 569-584.

Chalmers, D. and Bourget, D. (2021). *PhilPapers Survey 2020*. [online] Available at: https://survey2020.philpeople.org [Accessed 2 January 2022].

Cherniss, J. and Hardy, H. (2004). Isaiah Berlin. *Stanford Encyclopedia of Philosophy*. [online] Available at: https://plato.stanford.edu/archives/fall2020/entries/berlin [Accessed 4 April 2021].

Chomsky, N. and Segantini, T. (2015). History Doesn't Go In a Straight Line. *Jacobin Magazine*. [online] 22 September. Available at: https://chomsky.info/20150922/ [Accessed 21 April 2021].

Cohen, P.J. (1963). The Independence of the Continuum Hypothesis. *Proceedings of the National Academy of Sciences*, 50(6), pp. 1143-1148.

Cusanus, N. (1985). *Nicholas of Cusa On Learned Ignorance*. Translated by J. Hopkins. Minneapolis: A.J. Benning Press.

Da Costa, N.C.A. (2006). The Logic of Complementarity. In: J. Van Benthem, G. Heinzmann, M. Rebuschi and H. Visser, eds., *The Age of Alternative Logics*. Dordrecht: Springer.

De Broglie, L. (1923). Waves and Quanta. *Nature*, 112(2815), pp. 540-540.

De Saussure, F. (1959). *Course in General Linguistics*. Translated by W. Baskin. New York: Philosophical Library.

Deleuze, G. (2004). How Do We Recognize Structuralism? Translated by M. Taormina. In: D. Lapoujade, ed., *Desert Islands: and Other Texts, 1953–1974*. Los Angeles: Semiotext.

Descartes, R. (1971). Part Four. Translated by F.E. Sutcliffe. In: *Discourse on Method and the Meditations*. Harmondsworth: Penguin Classics.

Dietrich, E. (2011). There Is No Progress in Philosophy. *Essays in Philosophy*, 12(2), pp. 330-345.

Drucker, T., ed. (1991). *Perspectives on the History of Mathematical Logic*. Boston: Birkhäuser.

Dummett, M. (1995). *The Logical Basis of Metaphysics*. London: Duckworth.

Einstein, A. and Infeld, L. (1938). *The Evolution of Physics*. Cambridge: Cambridge University Press.

Ellerman, D.P. (1988). Category theory and concrete universals. *Erkenntnis*, 28(3), pp. 409-429.

Eppinger, H. and Hess, L. (1915). *Vagotonia: A Clinical Study in Vegetative Neurology*. New York: Nervous And Mental Disease Publishing Company.

Eysenck, H.J. (2014). *The Structure of Human Personality*. London; New York: Routledge.

Fichte, J.G. (1982). *The Science of Knowledge*. Translated by P. Heath and J. Lachs. Cambridge: Cambridge University Press.

Field, H. (2016). *Science Without Numbers*. Oxford: Oxford University Press.

Filk, T. (2014). Quantum Entanglement, Hidden Variables, and Acausal Correlations. In: H. Atmanspacher and C.A. Fuchs, eds., *The Pauli-Jung Conjecture and its Impact Today*. Exeter: Imprint Academic, pp. 109-123.

Firth, R. (1952). Ethical Absolutism and the Ideal Observer. *Philosophy and Phenomenological Research*, 12(3), pp. 317-345.

Fogelin, R.J. (1968). Wittgenstein and Intuitionism. *American Philosophical Quarterly*, 5(4), pp. 267-274.

Frege, G., Geach, P.T. and Black, M. (1951). On Concept and Object. *Mind*, 60(238), pp. 168-180.

Frege, G. (1980). *The Foundations of Arithmetic*. New York: Harper & Brothers.

Frege, G. (1990). Begriffsschrift (1879). In: J.V. Heijenoort, ed., *From Frege to Gödel*. Cambridge, MA; London: Harvard University Press.

Frege, G. (1999). *The Foundations of Arithmetic*. Evanston: Northwestern University Press.

Frege, G. (2013). Afterword. Translated by P.A. Ebert and M. Rossberg. In: *Basic Laws of Arithmetic*. Oxford: Oxford University Press.

Gao, S. (2011). *Why is Hume's Principle not good enough for Frege?* [online] Available at: http://www.cs.cmu.edu/~sicung/papers/frege.pdf [Accessed 21 July 2021].

Gazzaniga, M.S., Bogen, J.E. and Sperry, R.W. (1965). Observations of visual perception after disconnexion of the cerebral hemispheres in man. *Brain*, 88(B2), pp. 221-236.

Gödel, K. (1940). *The Consistency of the Continuum Hypothesis*. Princeton: Princeton University Press.

Gödel, K. (1990). On Formally Undecidable Propositions of Principia Mathematica and Related Systems (1931). In: J.V. Heijenoort, ed., *From Frege to Gödel*. Cambridge, MA; London: Harvard University Press.

Gödel, K. (1995). The modern development of the foundations of mathematics in the light of philosophy (1961). In: S. Feferman, W. Goldfarb, C. Parsons and R.N. Solovay, eds., *Collected Works (Volume III): Unpublished Essays and Lectures*. Oxford; New York: Oxford University Press.

Graham, D.W. (2007). Heraclitus. *Stanford Encyclopedia of Philosophy*. [online] Available at: https://plato.stanford.edu/entries/heraclitus [Accessed 4 April 2021].

Hayek, F. (2007). *The Road to Serfdom*. London; New York: Routledge Classics.

Heck, R.G. (1997). The Julius Caesar Objection. In: *Language, Thought, and Logic: Essays in Honour of Michael Dummett*. Oxford; New York: Oxford University Press.

Hegel, G.W.F. (1892). *The Logic of Hegel*. Translated by W. Wallace. Oxford: Clarendon Press.

Hegel, G.W.F. (1991). *Elements of the Philosophy of Right*. Translated by H.B. Nisbet. Cambridge; New York: Cambridge

University Press.

Hegel, G.W.F. (1997). *On Art, Religion, and the History of Philosophy: Introductory Lectures*. Indianapolis; Cambridge: Hackett Publishing Company.

Hegel, G.W.F. (2010). *The Science of Logic*. Translated by G. Di Giovanni. Cambridge; New York: Cambridge University Press.

Heisenberg, W. (1971). *Physics and Beyond. Encounters and Conversations*. London: Allen And Unwin.

Hernández-Tello, A., Macías, V.B. and Coniglio, M.E. (2020). Paracomplete Logics Dual to the Genuine Paraconsistent Logics: The Three-valued Case. *Electronic Notes in Theoretical Computer Science*, 354, pp. 61-74.

Hilbert, D. (1902). Mathematical Problems. *Bulletin of the American Mathematical Society*, 8(10), pp. 437-480.

Hilbert, D. (1970). Axiomatic Thinking. *Philosophia Mathematica*, s1-7(1-2), pp. 1-12.

Hofstadter, D.R. (1999). *Gödel, Escher, Bach: an Eternal Golden Braid*. New York: Basic Books.

Hofstadter, D.R. (2007). *I Am a Strange Loop*. London; New York: Basic Books.

Holton, G.J. (1970). The Roots of Complementarity. In: *The Making of Modern Science: Biographical Studies*. Cambridge: American Academy of Arts and Sciences, pp. 1015-1055.

Hume, D. (1985). *A Treatise of Human Nature*. London; New York: Penguin Classic.

Hume, D. (2008a). Section IV. In: P. Millican, ed., *An Enquiry Concerning Human Understanding*. Oxford; New York: Oxford University Press.

Hume, D. (2008b). Section VII, Part I. In: P. Millican, ed., *An Enquiry Concerning Human Understanding*. Oxford; New York: Oxford University Press.

Husserl, E. (2006). Investigation III: On the Theory of Wholes and Parts. Translated by J.N. Findlay. In: *Logical Investigations*

Volume 2. London; New York: Routledge.

Iyer, R., Koleva, S., Graham, J., Ditto, P. and Haidt, J. (2012). Understanding Libertarian Morality: The Psychological Dispositions of Self-Identified Libertarians. *PLoS ONE,* [online] 7(8). Available at: https://doi.org/10.1371/journal.pone.0042366 [Accessed 2 April 2020].

James, W. (1890). *The Principles of Psychology*. New York: Henry Holt and Company.

James, W. (1909). Lecture V. The Compounding of Consciousness. In: *A Pluralistic Universe*. London: Longmans, Green, and Co.

James, W. (1982). *The Varieties of Religious Experience: A Study in Human Nature*. Harmondsworth; New York: Penguin Classics.

James, W. (2000). *Pragmatism and Other Writings*. London; New York: Penguin Classics.

Jung, C.G. (1968). *Collected Works of C.G. Jung, Volume 12: Psychology and Alchemy*. Princeton: Princeton University Press.

Jung, C.G. (1973). *Collected Works of C.G. Jung, Volume 8: Structure & Dynamics of the Psyche*. Princeton: Princeton University Press.

Jung, C.G. (1985). *Practice of Psychotherapy*. 2nd ed. Princeton: Princeton University Press.

Jung, C.G. (1989). Septem Sermones ad Mortuos. Translated by R. Winston. In: A. Jaffe, ed., *Memories, Dreams, Reflections*. New York: Vintage Books.

Jung, C.G. (1992). *Letters*. London; New York: Routledge.

Jung, C.G. (2016). *Psychological Types*. London; New York: Routledge Classics.

Kabay, P.D. (2008). A defense of trivialism. PhD thesis, School of Philosophy, Anthropology, and Social Inquiry, The University of Melbourne. [online] Available at: https://minerva-access.unimelb.edu.au/handle/11343/35203 [Accessed 4 November 2021].

Kant, I. (1998). *Groundwork of the Metaphysics of Morals*. Translated by M. Gregor. Cambridge; New York: Cambridge University Press.

Kant, I. (2004). *Metaphysical Foundations of Natural Science*. Translated by M. Friedman. Cambridge; New York: Cambridge University Press.

Kant, I. (2007). *Critique of Pure Reason*. Translated by M. Weigelt. London: Penguin Classics.

Kneale, W. and Kneale, M. (1962). *The Development of Logic*. Oxford: Clarendon Press.

Kripke, S. (1975). Outline of a Theory of Truth. *The Journal of Philosophy*, 72(19).

Kristeva, J. (1982). *Powers of Horror: An Essay on Abjection*. New York: Columbia University Press.

Lacan, J. (1991). *The Ego in Freud's Theory and in the Technique of Psychoanalysis, 1954-1955 (The Seminar of Jacques Lacan: Book II)*. Translated by S. Tomaselli. New York; London: W.W. Norton & Company.

Lenin, V. (2021). *The Proletarian Revolution and the Renegade Kautsky*. Paris: Foreign Languages Press.

Lester, D. (1983). Constructing a questionnaire measure of autonomic nervous system balance. *Research Communications in Psychology, Psychiatry and Behavior*, 8(4), pp. 353-356.

Lester, D. and Berry, D. (1998). Autonomic Nervous System Balance and Introversion. *Perceptual and Motor Skills*, 87(3), pp. 882-882.

Levine, S. (2018). James and Phenomenology. In: A. Klein, ed., *The Oxford Handbook of William James*. Oxford; New York: Oxford University Press.

Lewis, D.K. (1986). *On the Plurality of Worlds*. Malden, MA: Blackwell Publishers.

Lienhard, D.A. (2017). Roger Sperry's Split Brain Experiments (1959–1968). *Embryo Project Encyclopedia*, [online] ISSN: 1940-5030. Available at: http://embryo.asu.edu/handle/10776/13035

[Accessed 5 January 2020].

Locke, J. (1997). Book II, Chapters 1 & 2. In: *An Essay Concerning Human Understanding*. London; New York: Penguin Classics.

Loux, M.J. (2017). Chapters 3 & 4. In: T.M. Crisp, ed., *Metaphysics: A Contemporary Introduction*. London; New York: Routledge.

Lucas, J.R. (1961). Minds, Machines and Gödel. *Philosophy*, 36(137).

MacIntyre, A. (2007). *After Virtue*. Notre Dame: University of Notre Dame Press.

Mansfield, V. (1991). The Opposites in Quantum Physics and Jungian Psychology. *Journal of Analytical Psychology*, 36(3), pp. 289-306.

Marks-Tarlow, T. (2013). Fractal Geometry as a Bridge between Realms. In: N. Sala and F. Orsucci, eds., *Complexity Science, Living Systems, and Reflexing Interfaces: New Models and Perspectives*. Hershey: IGI Global.

Marx, K. (1966). *Critique of the Gotha Program*. New York: International Publishers.

Maxwell, J.C. (1865). A Dynamical Theory of the Electromagnetic Field. *Philosophical Transactions of the Royal Society of London*, 155, pp. 459-512.

Mazzocchi, F. (2010). Complementarity in biology. *EMBO reports*, 11(5), pp. 339-344.

Mill, J.S. (1884). *A System of Logic: Ratiocinative and Inductive*. New York: Harper & Brothers.

Minsky, R. (1996). *Psychoanalysis and Gender: An Introductory Reader*. London; New York: Routledge.

Moore, G.E. (1903). *Principia Ethica*. London: Cambridge University Press.

Mouffe, C. (2000). *The Democratic Paradox*. London: Verso.

Mulligan, K., Simons, P. and Smith, B. (1984). Truth-Makers. *Philosophy and Phenomenological Research*, 44(3), pp. 287-321.

Nagel, T. (1989). *The View From Nowhere*. Oxford; New York: Oxford University Press.

Nelson, L. and Grelling, K. (1908). Bemerkungen Zu den Paradoxien von Russell Und Burali-Forti. *Abhandlungen der Fries'Schen Schule. Neue Folge*, 2, pp. 301-334.

Nozick, R. (1974). *Anarchy, State, and Utopia*. New York: Basic Books.

Osherson, D., Perani, D., Cappa, S., Schnur, T., Grassi, F. and Fazio, F. (1998). Distinct brain loci in deductive versus probabilistic reasoning. *Neuropsychologia*, 36(4), pp. 369-376.

Palmquist, S. (1993a). Knowledge and Experience. In: *Kant's System of Perspectives*. Lanham: University Press of America.

Palmquist, S. (1993b). Some Post-Kantian Variations of the Analytic-Synthetic Distinction. In: *Kant's System of Perspectives*. Lanham: University Press of America.

Parsons, L.M. (2001). New Evidence for Distinct Right and Left Brain Systems for Deductive versus Probabilistic Reasoning. *Cerebral Cortex*, 11(10), pp. 954-965.

Pauli, W. (1994). *Writings on Physics and Philosophy*. Translated by R. Schlapp. Berlin; New York: Springer.

Plato (1956). *Protagoras*. Translated by B. Jowett. New York: The Liberal Arts Press.

Plato (1993). *The Last Days of Socrates*. Translated by H. Tredennick. London: Penguin Books, p. 17, par.10a.

Plato (2005). *Protagoras and Meno*. Translated by A. Beresford. London; New York: Penguin Classics.

Plato (2009). *Phaedo*. Translated by D. Gallop. Oxford; New York: Oxford University Press.

Plato (2012). *Republic*. Translated by C. Rowe. London; New York: Penguin Classics.

Popper, K.R. and Eccles, J.C. (1985). *The Self and Its Brain*. Berlin: Springer International.

Priest, G. (1979). The Logic of Paradox. *Journal of Philosophical Logic*, 8(1).

Priest, G. (1994). The Structure of the Paradoxes of Self-Reference. *Mind*, 103(409), pp. 25-34.

Priest, G. (2000). On the Principle of Uniform Solution: a Reply to Smith. *Mind,* 109(433), pp. 123-126.

Priest, G. (2006). *In Contradiction*. Oxford; New York: Oxford University Press.

Priest, G. (2010). The Logic of the Catuskoti. *Comparative Philosophy,* 1(2), pp. 24-54.

Putnam, H. (2010). *Philosophy of Logic*. London; New York: Routledge.

Putnam, H. (2011). The Gödel Theorem and Human Nature. In: M. Baaz, ed., *Kurt Gödel and the Foundations of Mathematics: Horizons of Truth*. New York: Cambridge University Press.

Rabinowicz, W. and Rønnow-Rasmussen, T. (2003). Tropic of Value. *Philosophy and Phenomenological Research,* 66(2).

Rauscher, F. (2002). Kant's Moral Anti-Realism. *Journal of the History of Philosophy,* 40(4), pp. 477-499.

Rosen, S. (2014). Pauli's Dream: Jung, Modern Physics and Alchemy. *Quadrant,* 44(2).

Rothbard, M. (1975). Society Without A State. In: *Libertarian Forum*. [online] p. 3. Available at: https://www.rothbard.it/essays/society-without-a-state.pdf [Accessed 2 June 2020].

Russell, B. (1900). *A Critical Exposition of the Philosophy of Leibniz*. Cambridge: The University Press.

Russell, B. (1919). On Propositions: What They are and How They Mean. *Aristotelian Society Supplementary Volume,* 2(1).

Russell, B. (1990a). Letter to Frege (1902). In: J.V. Heijenoort, ed., *From Frege to Gödel*. Cambridge, MA; London: Harvard University Press.

Russell, B. (1990b). Mathematical logic as based on the theory of types (1908). In: J.V. Heijenoort, ed., *From Frege to Gödel*. Cambridge, MA; London: Harvard University Press.

Russell, B. (1999). *A History of Western Philosophy*. New York: Simon & Schuster.

Schopenhauer, A. (1909). *The World as Will and Idea (Vol. 2 of 3)*. Translated by R.B.H. Haldane and by J. Kemp. London:

Trench, Trübner & Co.

Schwartz, P. (1990). Libertarianism: the Perversion of Liberty. In: L. Peikoff, ed., *The Voice of Reason: Essays in Objectivist Thought*. New York: Meridian Books.

Sedley, D.N. (2007). *Plato's Cratylus*. Cambridge; New York: Cambridge University Press.

Semrud-Clikeman, M., Fine, J.G. and Zhu, D.C. (2011). The Role of the Right Hemisphere for Processing of Social Interactions in Normal Adults Using Functional Magnetic Resonance Imaging. *Neuropsychobiology*, 64(1), pp. 47-51.

Sextus Empiricus (1935). *Against the Logicians*. Translated by R.G. Bury. London: Heinemann.

Sextus Empiricus (1990). What Scepticism Is. Translated by R.G. Bury. In: *Outlines of Pyrrhonism*. Buffalo: Prometheus Books.

Shanker, S.G. (1987). The Nature of Proof. In: *Wittgenstein and the Turning-Point in the Philosophy of Mathematics*. Albany: State University of New York Press.

Smit, H. (2000). Kant on Marks and the Immediacy of Intuition. *Philosophical Review*, 109(2).

Smith, A. (1776). *An Inquiry into the Nature and Causes of the Wealth of Nations*. London: W. Strahan and T. Cadell.

Stern, R. (2007). Hegel, British Idealism, and the Curious Case of the Concrete Universal. *British Journal for the History of Philosophy*, 15(1).

Stone, M.H. (1936). The Theory of Representation for Boolean Algebras. *Transactions of the American Mathematical Society*, 40(1).

Tarski, A. (1956). The Concept of Truth in Formalized Languages. Translated by J.H. Woodger. In: *Logic, Semantics, Metamathematics: Papers From 1923 To 1938*. Oxford: Clarendon Press.

Tegmark, M. (2014). *Our Mathematical Universe: My Quest for the Ultimate Nature of Reality*. New York: Vintage Books.

Venn, J. (1880). On the diagrammatic and mechanical

representation of propositions and reasonings. *The London, Edinburgh, and Dublin Philosophical Magazine and Journal of Science*, 10(59), pp. 1-18.

Von Franz, M. (2001). *Psyche and Matter*. Shambhala Publications.

Von Neumann, J. (1958). *The Computer and the Brain*. New Haven: Yale University.

Wang, H. (2016). *A Logical Journey: From Gödel to Philosophy*. Cambridge, MA; London: MIT Press.

West, R. (1990). Relativism, Objectivity, and Law. *The Yale Law Journal*, 99(6).

Williams, D.C. (1953). On the Elements of Being: I. *The Review of Metaphysics*, 7(1), pp. 3-18.

Wittgenstein, L. (1978). *Remarks on the Foundations of Mathematics*. Translated by G.E.M. Anscombe. Oxford: Blackwell.

Wittgenstein, L. (1989). Lecture XXI. In: *Wittgenstein's Lectures on the Foundations of Mathematics, Cambridge, 1939*. Chicago: University of Chicago Press.

Wong, D.B. (1995). Pluralistic Relativism. *Midwest Studies in Philosophy*, 20, pp. 378-399.

Wong, D.B. (2019). Moral Ambivalence. In: M. Kusch, ed., *The Routledge Handbook of Philosophy of Relativism*. London; New York: Routledge.

Žižek, S. (2013). *Less Than Nothing: Hegel and the Shadow of Dialectical Materialism*. London: Verso Books.

Index

Abjection 60-61, 280
Abjectivism 60-61, 81, 143-148, 153, 211-213, 265-270, 280-281, 306-308, 326, 335-336
Absolute, The 33, 108-109, 255, 302, 318-320, 333, 335, 336
Absolute value 112–115, 119-121, 246-250
Absolutism (moral) 114-117, 120-121, 126, 130, 134, 147, 246-250, 311-312
Abstractness 97-106
Abstract particulars 103-106, 119, 145, 173, 281, 283-284, 301
Abstract universals 99-101, 107-108, 114, 244, 274
Acatalepsy 264
Acetylcholine 64-65
Agnosticism 260
Algebraic logic 158-262
Analyticity 86-89, 93-95, 240-243
Analytic a posteriori 93-94, 240-244, 248, 255, 264, 269, 292-293, 296
Analytic a priori 88-90, 99-100
Analytic philosophy 21-22, 34, 37, 166
Analytic psychology 50, 255, 281, 308
Analytic rationalism 89
Anarchism 125, 128-129
Anarcho-capitalism 128-129, 252, 306, 315
Anarcho-socialism 132-135
Anti-mechanism 220-226
Anti-realism (moral) 111-112, 116-117, 246-248
Aposteriority 85-95, 240-243
Apriority 85-93
Archetype 52, 281-285
Aristotle 18-20, 33-34, 97, 155-165, 223
Ataraxia 266
Atomism 75, 102-103
Authoritarian capitalism 130-132
Authoritarian collectivism 129-130
Authoritarian individualism 130-133, 135, 251, 308, 312
Authoritarian socialism 129-130
Authoritarianism 123-132
Autonomic nervous system 63-64
Awodey, Steve 279
Axiom of choice 180
Axiom schema of unrestricted comprehension 170
Axiomatic set theory 179-181, 291
Basic Law V 168-176, 180, 195-196
Beall, Jc 217, 261-262, 267
Benziger, Katherine 68-69, 95
Berlin, Isaiah 118, 125, 131, 251,

311
Berto, Francesco 192-194
Bidoctrinalism 311-317
Blackbody radiation 39
Bohr, Niels 37, 42-47, 260-261, 264, 324
Boole, George 158-162, 164
Boolean logic 160-161, 165-167
Bourbaki group 278
Brouwer, L. E. J. 191, 199
Brunner, A.B.M. 199
Buddhism 263-264, 266
Calculus 75
Calculus rationator 164-166
Cantor, Georg 76, 188
Capitalism 126-132
Carnielli, Walter 198-199
Categorical imperative 115, 247-249, 269
Category theory 278-279
Central nervous system 69-70
Characteristica universalis 164-165
Chihara, Charles 102
Chomsky, Noam 132
Church-Turing thesis 220
Classical Marxism 132-133
Cogito 89, 242
Cognitivism (moral) 111-113
Coincidentia oppositorum 50, 256
Collective unconscious 52, 281-283
Collectivism 125-135, 250-254
Communism 129, 132-133
Complementarity 34-36, 41-48, 67-69, 73-84, 93-94, 148-149, 198-204, 254, 260-261, 282-283, 306-308, 311, 323-323, 329
Completeness (logic) 182, 189, 194, 198, 203-204, 209, 233
Computation 160-161, 220-222, 232
Concreteness 97-98, 101-103, 106-109, 243-246
Concrete particular 101-103
Concrete universal 106-109, 243-246, 250, 253-255, 275, 288-294, 318-320
Consequentialism 76, 119, 128
Conservatism 124, 131, 251, 311-313
Consistency (logic) 182, 186-189, 193-195, 198, 203-204, 209, 214, 233
Constructivism (logic) 191
Continental philosophy 21, 34
Contingent a priori 243
Continuity 37, 40-41, 73-78, 96-98
Continuum hypothesis 76, 188, 192, 195, 200, 223
Coolness 337
Critical rationalism 143
Cumulative hierarchy 180, 278
Cusa, Nicholas of 50, 266
Da Costa, Newton 261
Daoism 47
De Morgan duality 213, 267

De Saussure, Ferdinand 274-275, 279-280, 310
Decidability 183-195, 200-203, 220-223
Deduction 95, 143, 292
Deflationary theory of truth 262
Deleuze, Gilles 291
Democracy 134-135
Democratic paradox 251
Democritus 75, 120
Descartes, René 25, 37, 89, 242
Determinateness 78-82, 87, 122, 141, 255
Developmental psychology 319
Dialectic (Hegel) 31-36, 108, 253-254, 320
Dialetheia 196-197, 278
Dialetheism 192-197
Dialogic 36, 254
Dietrich, Eric 14
Discreteness 40-41, 73-76
Discursive mark 99
Divine command theory 121, 250
Divine right of kings 130
Doctrine of two truths 265
Dopamine 64-65
Dostoevsky, Fyodor 125
Double negation 32, 192, 213, 218, 267
Double-slit experiment 38-39
Dual-aspect monism 282, 305
Duality 14-15, 25, 302-308, 326-329
Duality (logic) 171, 198-204, 267
Dummett, Michael 199
Egoism 126, 129
Einstein, Albert 40-42
Electromagnetism 39-40, 77
Eleaticism 75
Eliminative materialism 102-103
Emergent materialism 103, 230-231
Emotivism 116
Empiricism 27-28, 49, 89-91
Empty set 160, 180
Enlightenment
Enlightenment (period) 37, 97, 119, 124
Essential dialectic 264-272, 282
Genetics 64, 229, 323
Epimenides 178
Epoché 335
Eppinger, Hans 63
Equality 127-135, 251-254, 306, 312-317
Ethical life (Hegel) 253
Eubulides 178
Euthyphro dilemma 250
Evolution (biological) 17, 22, 25, 52, 67, 300-301, 320-329
Extraversion 51-60, 63-65
Extraverted feeling 58, 60
Extraverted thinking 56
Extremism 306-308
Eysenck, Hans 63
Facetiousness 313-315
Falsification 143, 199-203, 207-208, 262, 274-275

Falsification of type 68
Feeling (typology) 53-61, 65-69
Fibonacci sequence 285
Fichte, Johann 232, 242, 320
Field, Hartry 102
Filk, Thomas 305
Finalism 323-324
Firth, Roderick 120
Flux 20, 79, 269
Form of the Good 78, 114, 249
Forms (Platonic) 96-101, 114, 249
Formal dialectic 264-272, 282
Formalism (logic) 189-191, 210, 222
Fractals 287-289
Frege, Gottlob 88-89, 97, 165-182, 195-196
Frege's theorem 170-172
Gao, Sircun 173
Gaps 202, 213-214, 259, 262-264, 267, 272
Glanzberg, Michael 217
Gluts 202-203, 213-214, 217-218, 259, 262-264, 272
Gödel 182-194, 198-202, 214-226, 229-231, 277-278, 300-302
Gödel coding 184-186, 216, 229, 277
Gödel number 184-186, 229
Gödel's incompleteness theorems 182-194, 198, 214, 220-224
Golden rule 115, 247
Grelling, Kurt 177
Grelling-Nelson paradox 177-178
Halting problem 220
Hayek, Friedrich 128-129
Hegel, Georg 31-37, 44, 83, 107-108, 232, 244-245, 253, 268, 319-320
Heisenberg, Werner 42-43, 260
Heraclitus 20, 79, 268-269
Hertz, Heinrich 39
Hess, Leo 63
Hilbert's program 182-183, 186-189
Hilbert, David 182-183, 186-189, 194, 222
Hobbes, Thomas 126
Hoftstadter, Douglas 176-177, 225-233, 258, 278
Homotopy type theory 278-279
Hume, David 27-29, 92, 116, 170
Hume's fork 92
Hume's principle 170-173
Humean scepticism 27-29, 105
Husserl, Edmund 104, 301-302
Huygens, Christiaan 38
Ideal observer theory 120-121, 147, 250
Idealism 99-103
Ideologism 89
Imaginary numbers 285-287
Immutability 79-82, 87, 122, 141, 255
Impassioned indifference 335-338
Individualism 125-132
Individuality 107-108, 244, 251, 255, 263, 318-319

Individuation 315, 319
Induction 27, 95, 143-144
Intaglio sculpture 201-204, 272, 310
Integration 75, 286
Intentionality 302
Interference pattern 38, 77
Introversion 51-60, 63-65
Introverted feeling 57-58
Introverted thinking 56-59
Intuition (typology) 53, 57
Intuitionism 189-192, 199, 210
Intuitive mark 105
Invariance (physics) 92
Isomorphism 140-141, 160, 258, 275, 277-279, 288, 298
James, William 46-51, 54-55, 77, 298, 300, 302
Jung, Carl 50-64, 67-69, 89, 146, 255-256, 266-267, 281-288, 290, 301, 304-305, 308, 319, 325
Kabay, Paul 266
Kant, Immanuel 28-34, 44-45, 83, 85-87, 90-94, 99-100, 105-106, 115, 241-242, 247-249, 269, 301
Kant's antinomies 92
Kitcher, Philip 243
Kleene, Stephen Cole 217-218
Kneale, William and Martha 161
Kripke, Saul 217, 243
Kripke-Kleene models 217
Kristeva, Julia 61, 280
Lacan, Jacques 280-281, 290-291
Laozi 19
Lateralisation of brain function 65-70
Law of excluded middle 156, 159-161, 183, 197-198, 203, 209-214, 266
Law of identity 156, 159-160, 232
Law of non-contradiction 19-20, 34, 155-156, 161, 195-198, 209-214, 266
Laws of thought 153
Leibniz, Gottfried 75, 89, 163-165, 304
Lenin, Vladimir 129, 133
Leninism 133
Lester, David 63-64
Lucretius 102
Liar paradox 178, 185, 216-217, 226
Liberalism 131, 251
Libertarian capitalism 128-129
Libertarian collectivism 132-135, 250-255, 316-317
Libertarian individualism 128-129
Libertarian socialism 132-135
Libertarianism 123-129, 132-135, 250-254
Liberty 123-125, 128-135, 251-254, 311-317
Linguistics 274-275, 280
Locke, John 28, 90
Logic of indeterminacy 218
Logic of paradox 218
Logical pluralism 260-262

Logicism 90, 165-175, 191, 195
Logos 169
Lucas, John 223-226
Luck (chance) 324-326
MacIntyre, Alasdair 119
Mandelbrot set 287
Marks-Tarlow, Terry 287-288
Malebranche, Nicolas 304
Marx, Karl 129, 253-254
Marxism 129, 132-135, 253-254
Mathematical empiricism 90
Mathematical universe hypothesis 106
Matrices 137-140
Maxwell, James Clerk 39-40
Myers-Briggs Type Indicator 69
Measurement 77-79, 305
Mechanism (philosophy) 22, 45, 220-233, 323-326
Meinong, Alexius 275
Mendeleev, Dmitri 139
Meritocracy 129
Metalanguage 216-218
Metaphilosophy 237-239
Metatheory 189-193, 216, 221-222
Mill, John Stuart 90-91, 128
Miller, David 199
Miller, Jacque-Alain 291
Millikan, Robert 41
Minarchism 128
Moderatism 306-308, 311-312
Moments (Husserl) 301
Monoletheism 16-19, 31-33, 83-84, 153-154, 203, 218, 219, 261, 307
Moore, G.E. 34, 114
Mouffe, Chantal 251, 254, 265-266
Nāgārjuna 19, 336
Nagel, Thomas 14
Naive realism 25-28, 105
Naive set theory 165-174
Natural philosophy 18
Natural selection 17, 52, 323-324
Naturalism 18, 27, 96, 103, 114
Nature of light 37-46
Negative facts 273, 274
Negative liberty 124-125
Negativism 143
Nelson, Leonard 177
Neo-Darwinism 323
Neuroses 51, 62-63
Neutral monism 106, 146
Newton, Isaac 38-39, 75
Newtonian mechanics 37, 92
Ninth view (emptiness) 336
Nominalism 102-104, 110, 119
Non-classical logic 191-219, 223, 260-263
Non-cognitivism 113, 116
Nonduality 264, 294, 327, 336
Non-naturalism (ethics) 112, 115, 148
Normative ethics 111-121, 147, 246, 311-312
Nothing (logic) 160
Nothingness 2-4, 263-273, 280,

290-291
Noumenal 29, 33, 45, 83, 279-281, 301-302
Numbers 96-97, 106, 168-169, 173, 179, 284-288
Objective absolute value 113-115, 130
Objective idealism 101, 103
Objective relative value 117-119, 130
Objective value 110-119
Objectivism (syntheorology) 58, 77, 111, 143-144, 147-148, 209, 265, 270
Open-question argument 114
Optimism 49, 133, 265, 301
Palmquist, Stephen 241-243, 249, 269
Paracomplete logic 197-212, 217-218, 227, 258-262
Paraconsistent logic 198-212, 217-218, 227, 258-262, 273-275, 278
Paradox 170-185, 193-196, 202-204, 212 -213, 216-219, 226-228, 238-239, 258-263, 277-278, 319-320, 331-333
Paraphilosophy 275-277, 292-317, 330, 335-336
Parasympathetic nervous system 63-64
Parmenides 79, 96
Particularity 97-98, 101-106
Pauli, Wolfgang 282-283, 286, 305, 324-325
Peano axioms 169-170, 180
Penrose, Roger 223
Periodic table 139, 293
Peripheral nervous system 63-64
Personal unconscious 281
Personality 63, 68
Pessimism 49, 133, 135, 254, 265, 301
Phantasiai 264
Phenomenology 104, 301-302, 318-329
Photoelectric effect 40-41, 77
Physicalism 18, 25, 103, 231-233
Physikoi 18
Planck, Max 40
Plato 18, 31, 78, 96-104, 114, 249-250, 327
Platonic realism 101-104
Platonic solids 137
Pleroma 267, 290
Popper, Karl 143, 274
Positive liberty 125
Positivism 37, 144, 183, 187, 199
Pragmatism 50
Predicate logic 165-169
Prefrontal cortex 68
Presocratic philosophy 20, 75, 96
Pribram, Karl 68
Priest, Graham 177, 193, 195-196, 218, 263, 277-278
Principle of explosion 172, 196-198, 209-212, 216
Principle of implosion 198, 209-

212
Principle of structuralism 279
Principle of uniform solution 177
Protagoras 116
Psychological type 54-62
Psychologism 89
Psychophysical parallelism 304-305
Putnam, Hilary 91, 102
Pyrrhonism 27, 264, 266, 335
Pythagoras 96, 284
Quantum mechanics 17, 40-44, 77, 261, 283-284, 305, 325
Quine, Willard 91, 102
Rayleigh-Jeans law 40
Randomness 324-325
Rationalism 49, 89, 93
Rawl, John 247-248
Realism (moral) 110-115, 117-119, 123-124, 247
Reductive physicalism 103, 306
Relative value 112-119
Relativism (moral) 112-113, 116-120, 126-130,
Relief sculpture 201-204, 272
Restall, Greg 261
Revolutionary spontaneity 133
Ripley, David 217
Romanticism 37, 49
Rothbard, Murray 128
Rousseau, Jean-Jacques 126
Russell, Bertrand 88-89, 170-180, 183, 193-195, 273, 278
Russell's paradox 170-174, 177, 181, 196, 277-278
Scepticism 27-29, 105, 266, 300, 335
Schopenhauer, Arthur 157
Schroder, Ernst 166
Second-order logic 182
Self 50-51, 230-232, 243-244, 255-256, 290-294, 297, 319-323, 327-330
Self-position 230-232, 243, 293, 320
Self-reference 171-172, 176-178, 183-185, 205, 213, 221, 227-229, 238-239, 244, 250, 258-259, 270, 292-294, 318-319
Self-similarity 287-292
Sensation (typology) 53, 56
Sentimentalism (moral) 117
Seriousness 313-315
Smit, Houston 105
Smith, Adam 126
Smith, Barbara 123-124
Socialism 129-135
Socrates 100, 250, 327
Sovereignty of the individual 252-253
Sperry, Roger 65-66
Spinoza, Baruch 55, 304
Spiritualism 57, 266, 332-334
Split-brain 65
Stern, Robert 107-108, 245
Strange loops 227-233, 239, 243, 250-251
Stratification 176-179

Structuralism 277-281
Subjective absolute value 119-121, 246-250
Subjective idealism 101
Subjective relative value 116-117
Subjective value 110-113, 116-121, 246-250
Subjectivism (syntheorology) 58, 77, 58, 77, 111, 143-144, 147-148, 209, 265, 270
Sublation 32-35
Substance dualism 25
Superego 61
Superjectivism 58-62, 81, 146-149, 212-214, 240-270
Syllogistic logic 154-158
Symbolic logic 158-165
Sympathetic nervous system 63-64
Synchronicity 305-306, 325
Syntheorology 82-83, 136-149, 277-282, 293, 301-311
Synthetic a posteriori 89-91
Synthetic a priori 91-93, 269
Synthetic rationalism 93
Syntheticity 86-93
Taijitu 47
Tarski, Alfred 214-219
Tarski's undefinability theorem 214-219
Tegmark, Max 284
Teleology 119, 253, 272, 323-325
Tender-mindedness 49, 77, 300
The archetype 281-285
The imaginary (psychoanalysis) 280
The real (psychoanalysis) 280-282
The symbolic (psychoanalysis) 280-281
Theory of recollection 100
Thinking (typology) 53-60, 65-69
Totalitarianism 130
Tough-mindedness 49, 77, 300
Transcendentalism 28-30, 283, 308
Translation invariance 92
Transparency of truth 262
Trivialism 196-198, 212, 216, 266-267
Trope theory 98, 104-105, 119, 301
Truth values 19, 161, 197, 209-213, 217-218
Turing, Alan 220
Type theory 175-179, 278
Typology (psychology) 50-70
Uncertainty principle 42-43
Unconscious 51-52, 63, 281-283, 286-288, 304-305
Unity of opposites 34-36, 239-256
unity-amid-diversity 268-269
Univalence axiom 279
Universal set 159, 171, 175
Universalsality 97-101, 106-109, 243-246
Universe of discourse 159-161, 181
Unus mundus 283, 305

Upekṣā 266
Value pluralism 118-119, 130-131, 311
Venn, John 158, 160
Verification 143-144, 187, 199-203, 207-209, 162, 275
Vicious circle principle 176-179, 184
Virtue ethics 119
Von Franz, Marie-Louise 284-286
Von Mises, Ludwig 128
Von Neumann, John 181
Wang, Hao 222
Wave-particle duality 41-44
Wheeler, John 43-44
Williams, Donald Cary 104
Wisdom 320-321, 327, 332-337
Wittgenstein, Ludwig 190-197
Wong, David 118
Yin and yang 47
Young, Thomas 38-39
Zeno of Elea 75
Zeno's paradoxes 75
Zermelo, Ernst 179-183
Zermelo set theory 179-183, 186, 188
Zero-energy universe hypothesis 282
ZFC 181
Žižek, Slavoj 35

ACADEMIC AND SPECIALIST

Iff Books publishes non-fiction. It aims to work with authors and titles that augment our understanding of the human condition, society and civilisation, and the world or universe in which we live.

If you have enjoyed this book, why not tell other readers by posting a review on your preferred book site.

Recent bestsellers from Iff Books are:

Why Materialism Is Baloney
How true skeptics know there is no death and fathom answers to life, the universe, and everything
Bernardo Kastrup
A hard-nosed, logical, and skeptic non-materialist metaphysics, according to which the body is in mind, not mind in the body.
Paperback: 978-1-78279-362-5 ebook: 978-1-78279-361-8

The Fall
Steve Taylor
The Fall discusses human achievement versus the issues of war, patriarchy and social inequality.
Paperback: 978-1-78535-804-3 ebook: 978-1-78535-805-0

Brief Peeks Beyond
Critical essays on metaphysics, neuroscience, free will, skepticism and culture
Bernardo Kastrup
An incisive, original, compelling alternative to current mainstream cultural views and assumptions.
Paperback: 978-1-78535-018-4 ebook: 978-1-78535-019-1

Framespotting
Changing how you look at things changes how you see them
Laurence & Alison Matthews
A punchy, upbeat guide to framespotting. Spot deceptions and hidden assumptions; swap growth for growing up. See and be free.
Paperback: 978-1-78279-689-3 ebook: 978-1-78279-822-4

Is There an Afterlife?
David Fontana
Is there an Afterlife? If so what is it like? How do Western ideas of the afterlife compare with Eastern? David Fontana presents the historical and contemporary evidence for survival of physical death.
Paperback: 978-1-90381-690-5

Nothing Matters
a book about nothing
Ronald Green
Thinking about Nothing opens the world to everything by illuminating new angles to old problems and stimulating new ways of thinking.
Paperback: 978-1-84694-707-0 ebook: 978-1-78099-016-3

Panpsychism
The Philosophy of the Sensuous Cosmos
Peter Ells
Are free will and mind chimeras? This book, anti-materialistic but respecting science, answers: No! Mind is foundational to all existence.
Paperback: 978-1-84694-505-2 ebook: 978-1-78099-018-7

Punk Science
Inside the Mind of God
Manjir Samanta-Laughton
Many have experienced unexplainable phenomena; God, psychic abilities, extraordinary healing and angelic encounters. Can cutting-edge science actually explain phenomena previously thought of as 'paranormal'?
Paperback: 978-1-90504-793-2

The Vagabond Spirit of Poetry

Edward Clarke

Spend time with the wisest poets of the modern age and of the past, and let Edward Clarke remind you of the importance of poetry in our industrialized world.

Paperback: 978-1-78279-370-0 ebook: 978-1-78279-369-4